SpringerBriefs in Population Studies

Advisory Board

Baha Abu-Laban, Edmonton, AB, Canada
Mark Birkin, Leeds, UK
Dudley L. Poston, Jr., Department of Sociology, Texas A&M University,
College Station, TX, USA
John Stillwell, Leeds, UK
Hans-Werner Wahl, Deutsche Zentrum für Alternsforschung (DZFA),
Institut für Gerontologie, Universität Heidelberg, Heidelberg, Germany
D. J. H. Deeg, VU University Medical Centre/LASA, Amsterdam, The Netherlands

SpringerBriefs in Population Studies presents concise summaries of cutting-edge research and practical applications across the field of demography and population studies. It publishes compact refereed monographs under the editorial supervision of an international Advisory Board. Volumes are compact, 50 to 125 pages, with a clear focus. The series covers a range of content from professional to academic such as: timely reports of state-of-the art analytical techniques, bridges between new research results, snapshots of hot and/or emerging topics, and in-depth case studies.

The scope of the series spans the entire field of demography and population studies, with a view to significantly advance research. The character of the series is international and multidisciplinary and includes research areas such as: population aging, fertility and family dynamics, demography, migration, population health, household structures, mortality, human geography and environment. Volumes in this series may analyze past, present and/or future trends, as well as their determinants and consequences. Both solicited and unsolicited manuscripts are considered for publication in this series.

SpringerBriefs in Population Studies will be of interest to a wide range of individuals with interests in population studies, including demographers, population geographers, sociologists, economists, political scientists, epidemiologists and health researchers as well as practitioners across the social sciences.

More information about this series at http://www.springer.com/series/10047

William P. O'Hare

Differential Undercounts in the U.S. Census

Who is Missed?

William P. O'Hare
O'Hare Data and Demographic Services
LLC
Cape Charles, VA, USA

ISSN 2211-3215   ISSN 2211-3223   (electronic)
SpringerBriefs in Population Studies
ISBN 978-3-030-10972-1   ISBN 978-3-030-10973-8   (eBook)
https://doi.org/10.1007/978-3-030-10973-8

Library of Congress Control Number: 2019930974

© The Editor(s) (if applicable) and The Author(s) 2019. This book is an open access publication.
Open Access This book is licensed under the terms of the Creative Commons Attribution 4.0 International License (http://creativecommons.org/licenses/by/4.0/), which permits use, sharing, adaptation, distribution and reproduction in any medium or format, as long as you give appropriate credit to the original author(s) and the source, provide a link to the Creative Commons license and indicate if changes were made.
The images or other third party material in this book are included in the book's Creative Commons license, unless indicated otherwise in a credit line to the material. If material is not included in the book's Creative Commons license and your intended use is not permitted by statutory regulation or exceeds the permitted use, you will need to obtain permission directly from the copyright holder.
The use of general descriptive names, registered names, trademarks, service marks, etc. in this publication does not imply, even in the absence of a specific statement, that such names are exempt from the relevant protective laws and regulations and therefore free for general use.
The publisher, the authors and the editors are safe to assume that the advice and information in this book are believed to be true and accurate at the date of publication. Neither the publisher nor the authors or the editors give a warranty, express or implied, with respect to the material contained herein or for any errors or omissions that may have been made. The publisher remains neutral with regard to jurisdictional claims in published maps and institutional affiliations.

This Springer imprint is published by the registered company Springer Nature Switzerland AG
The registered company address is: Gewerbestrasse 11, 6330 Cham, Switzerland

Acknowledgements

I would like to thank many current and former Census Bureau employees who have given me invaluable advice and feedback. These include Scott Konicki, Eric Jensen, Deborah Griffin, Howard Hogan, Heather King, Yeris Mayol-Garcia, Greg Robinson, and Mary Mulry. I would also like to thank Hannah Denny for editorial assistance. A special thanks to my wife (Barbara O'Hare) for providing extremely helpful feedback and valuable suggestions on earlier drafts of the book. And I would like to acknowledge support from the 2020 Census Project, which is a pooled fund administered by a Census funder collaborative promoting a fair and accurate Census. None of the people above are responsible for any shortcomings or errors in the publication.

Contents

Chapter 1
Who Is Missing? Undercounts and Omissions in the U.S. Census

Abstract Over the past 60 years, the overall accuracy of the U.S. Decennial Census has steadily improved. But some groups still experience higher net undercounts than other groups in the Census. The issue of differential Census undercounts is introduced in this Chapter along with some of the key concepts related to measuring the accuracy of Census counts, sometimes called Census coverage. Some of the key terminology is also discussed in this Chapter along with a description of the intended audience for this publication. The contents of the publication are described Chapter by Chapter.

1.1 Introduction

The mantra of the U.S. Census Bureau is to count every person once, only once, and in the right place. This is easy to say, but difficult to achieve. The U.S. Census Bureau tries very hard to include every person in the Decennial Census, but some people are always missed. The situation is summed up neatly by the U.S. General Accounting Office (2003, p. 4),

> The Bureau puts forth tremendous effort to conduct a complete and accurate count of the nation's population. However, some degree of error in the form of persons missed or counted more than once is inevitable because of limitations in Census-taking methods.

This thought is echoed by Raymondo (1992, p. 37),

> As most people know, the census of population is intended to count each and every resident of the United States. As most people might suspect, any undertaking so ambitious is bound to fall short to some degree, and the fact is that the census does not count each and every person. The failure to count everyone in the census is referred to as an undercount or, more generally as a coverage error.

The extent to which people in certain groups are missed or counted more than once is reflected in Census coverage measurements. The most widely used measures of Census coverage or Census accuracy are net undercounts, net overcounts and omissions.

What is a net undercount in the Census? In every demographic group, some people are missed in the Census, and some people are counted more than once (or included inappropriately) in the Census. When the number of people missed is larger than the

© The Author(s) 2019
W. P. O'Hare, *Differential Undercounts in the U.S. Census*,
SpringerBriefs in Population Studies, https://doi.org/10.1007/978-3-030-10973-8_1

number of people counted more than once, it produces a net undercount. When the number of people counted more than once is larger than the number of people missed it produces a net overcount. Census Bureau whole person imputations are another component in this equation, but for the sake of simplicity they are ignored for now.

Net undercounts have been measured often and consistently in the U.S. Census over the past 60 years, and it has been the main measure used by demographers to assess Census accuracy. Robinson (2010) refers to the net undercount from the Census Bureau's Demographic Analysis method as the "gold standard." A more detailed discussion of this topic is provided in Sect. 1.4 of this Chapter and in Chap. 3.

Omissions are a key component of net undercounts. Omissions reflect people who are missed in the Census and as such they are an important focus of this book. In some ways, omissions are a better measure of Census quality than net undercounts because double counting can mask omissions. A net undercount of zero could be the result of no one being missed and no one double counted, or for example, it could reflect a situation where ten percent of the population are missed, and ten percent are double counted. Differential omissions rates reflect that same kind of inequity as differential undercount rates. Data on the characteristics of people who are missed are presented in most of the Chapters in this book.

Unfortunately, people in some groups have higher net undercount rates than people in other groups. These differences in Census coverage are referred to as a "differential undercounts" and are the focus of this book. More specifically this book focuses on the groups that have the highest net undercounts and highest omissions rates in the U.S. Census.

The main point of this publication is to present existing information on net under-counts and omissions in the U.S. Census in a simple, organized, and systematic way. I have not found a single publication that pulls together key undercount and omissions data from different Census assessment methods, for many groups, over several times periods. This publication aims to fill that niche. By putting the key information into one publication I hope this publication makes key data on Census accuracy more readily available to a wider, non-technical audience. I also hope this publication facilitates further research on Census coverage issues.

By focusing on which groups have the highest net undercounts and Census cov-erage differentials the primary focus of the publication is descriptive rather than analytical. Material that tries to explain why net undercounts (or net overcounts) occur is more limited in this publication.

The book draws heavily on data produced by the U.S. Census Bureau. Most of the statistics presented here are publicly available, but the key data are often buried in large statistical reports, available on some obscure portion of the Census Bureau's website, appear only in scholarly journals, in presentations made at scientific con-ferences, or appear in internal Census Bureau reports (see, for example, U.S. Census Bureau 1974, 2012; Fay et al. 1988; Robinson et al. 1993; Robinson and Adlaka 2002; Mayol-Garcia and Robinson 2011). In some cases, one must download and analyze data files from the Census Bureau to produce simple net undercount tab-ulations. For a large share of the public such information is not easily or readily available.

More specifically, this report will draw extensively on data produced by the Census Bureau's Demographic Analysis (DA) and Dual-Systems Estimates (DSE) programs. More information about these two methods is provided in Chap. 3.

Although Demographic Analysis and Dual-Systems Estimates are the best estimates available regarding Census coverage it is important to recognize these methods have some limitations. For example, the demographic groups for which these two programs produce data are limited. Census Bureau tabulations have focused on measuring Census coverage for five demographic characteristics including;

- Age
- Sex
- Race
- Hispanic Origin Status
- Tenure

The 2010 DSE reports covers all these characteristics and the 2010 DA covers all but tenure, at least to some degree.

It should be noted that many times the most important differential undercounts are based on a combination of the characteristics noted above. For example, the net undercount for Black males age 30–49 is much higher than the net undercount rate for the total Black population or total male population. These special populations will be highlighted in the appropriate Chapters.

The characteristics noted above are important, but there are many other groups for which we would like to have data on Census undercounts. Many such groups are highlighted in public discussion of the Census. For example, in response to release of 2010 Census results, former Undersecretary of Commerce Dr. Rebecca Blank (2012, p. 1) said,

> However, as has been the case for some time, today's release shows that certain populations were undercounted. More work remains to address persistent causes of undercounting, such as poverty, mobility, language isolation, low levels of education, and general awareness of the survey.

Undercounts for the groups mentioned by Dr. Blank are not measured by the Census Bureau's DA or DSE methods. Some of these groups are captured by the "hard-to-count" factors (Bruce and Robinson 2003) and Mail Return Rates (Letourneau 2012) used by the Census Bureau and Census advocates. To be clear, this publication focuses on direct measures of Census coverage (net undercounts, net overcounts, and omissions) rather than metrics that reflect likelihood of being counted accurately like Mail Return Rates or hard-to-count scores.

In addition to reports and datasets from DA and DSE, I will add information from Census Bureau reports focused on some Census operations. Information on Census operations can shed light on the mechanisms by which differential Census undercounts occur. For example, see U.S. Census Bureau reports on topics such as Census Followup, Non-Response Followup, and Mail Return Rates (Govern et al. 2012; Letourneau 2012). I also draw on a series of qualitative studies that can help us understand why Census errors occur (de la Puente 1993; Schwede 2003, 2006; Schwede and Terry 2013).

The focus of this report is on Census net undercounts and omissions, but it should be noted that these are not the only types of Census errors. Census errors also include net overcounts and erroneous inclusions. Erroneous inclusions are people who are counted more than once and people who are included in the U.S. Census inappropriately. For example, an erroneous inclusion would be a foreign tourist who gets included in the Census by mistake or someone who dies before the Census date but is included inappropriately in the Census count. Another type of error is counting people in the wrong place.

From a scientific perspective net overcounts and erroneous inclusions are measurement errors just like net undercounts and omissions. But net undercounts and omissions are a much bigger public relations and public perception problem. According to Williams (2012, p. 8), "Differential undercounts are a recurrent problem in the Decennial Census and diminish the perception that the count is equitable to the entire population."

Net undercounts and omissions are much more of a problem than net overcounts. I am not aware of any lawsuit brought by a state or city because of real or perceived net overcounts, but there have been many lawsuits brought because of real or perceived net undercounts. Undercounts are also much more of a public relations problem for the Census Bureau. Note in the previous quote from Dr. Blank, she focuses on the problem of undercounts in the Census, not overcounts. Given this situation this book will focus on net undercounts and omissions with only passing note on net overcounts or erroneous inclusions.

1.2 Audience

This book is aimed largely at people outside the scholarly community such as practitioners and advocates. Because of the social equity issues raised by the differential net undercounts for many racial and Hispanic minority groups the book will be of interest to many civil rights organizations such as the Leadership Conference on Civil Rights, National Association of Latino Elected Officials, the Mexican-American Legal Defense and Education Fund, The National Urban League, The National Association for the Advancement of Colored People, Asian-American Advancing Justice League, National Congress of American Indians, and many others. One of the main purposes for writing this book is to make the high-quality data on Census errors available to a wider audience. Too many times, I have heard people assert that there is a net undercount for this group or that group when there is no good evidence to support that claim.

The book may also be useful for researchers in the demographic community because it fills an important niche regarding Census accuracy. It will provide a handy reference for the relative Census coverage rates for many key populations. In the context of scholarship, the book will help round out the literature in demography and/or population studies courses. Since the publication provides a lot of information in one place it could be a useful reference book for Census Bureau staff and related government organizations such as the U.S. General Accountability Office, The Congressional Research Service, and The U.S. Office of Management and Budget. Other possible users include professional organizations that monitor the Census such as the Population Association of America, the American Statistical Association, American Association of Public Opinion Researchers, and the non-profit organizations such as Population Reference Bureau, The Funders Census Initiative, and The Census Project.

Given this audience, some of the more detailed and esoteric aspects of the statistical methods used to assess Census coverage are ignored in favor of more straightforward language and a focus on results rather than methods. Readers who are more technically inclined can find the more detailed information through the citations offered in this publication, and the readers who are not technically inclined will get the basic information.

A significant segment of the audience for this book will be focused on one group or one Chapter. For example, the National Association for the Advancement of Colored People are likely to focus on the Chapter related to Blacks and the National Association for Latino Elected Officials are likely to focus on the Chapter related to Hispanics. Partnership for America's Children will be more interested in the Chapter on age which shows a high net undercount of young children.

1.3 Terminology

Some of the language, terminology, and nomenclature used in this publication may be unfamiliar to many readers and some terms have been used inappropriately or incorrectly in the past. In addition, some of the key terms may sound like the same thing but they have a different meaning to demographers. In general, I follow the nomenclature conventions of the U.S. Census Bureau.

The terms "net undercount" and "net overcount" have very precise meanings to demographers. But the terms "undercount" or "net undercount" are sometimes used loosely by non-demographers to mean people missed in the Census in a broad sense. In this publication the focus is on scientific measurements of net undercounts.

It is important to recognize that the net undercount does not reflect the number of people missed even though the term undercount is often used to suggest this. As stated earlier, net undercounts reflect a balance of people missed and people counted more than once or otherwise included erroneously. Demographers use the terms gross undercount or omissions to reflect the number of people missed.

Only the Dual-Systems Estimates (DSE) method produces data for omissions. I use omissions in many portions of this publication to supplement data on net undercounts. A more detailed discussion of methodology is offered in Chap. 3.

Prior to the 2010 Census, whenever the Census count was less than the DA estimate the Census Bureau typically reported the difference between a DA estimate and the

Census as an undercount. But some of the information put out by the Census Bureau following the 2010 Census refers to differences between the Census count and the DA estimates rather than net undercounts or net overcounts. This is meant to reflect the fact that both the Census and the estimate to which the Census is being compared (DA or DSE) have errors. While I understand the intent of using the term "differences" rather than undercount and overcounts, I will use the traditional terms net undercount and net overcount because these terms are more widely understood, and they indicate the directionality of the differences which makes communication more efficient. In other words, saying there is a one percent difference between the Census count and the Demographic Analysis (DA) estimate does not tell a reader if the Census is larger or smaller than the DA estimate, but saying there is a one percent net undercount indicates the Census count is lower than the DA estimate.

Another issue that might cause confusion is the fact that undercounts have sometimes been reported as a negative number by the Census Bureau (Velkoff 2011) and sometimes as a positive number by the Census Bureau (U.S. Census Bureau 2012). Since I draw on Census Bureau reports that use both expressions for net undercounts, I thought it important to standardize presentation within this publication. In the remainder of this publication, the differences between the Census counts and DA or DSE estimates are shown as the Census count minus the DA or DSE estimate. So, a negative number reflects a net undercount. This is consistent with the convention used by Velkoff (2011) in reporting the first results of the 2010 DA. This presentation style was also used in a couple of recent Census Bureau papers on this topic (King et al. 2018; Jensen et al. 2018). Also, this approach is consistent with O'Hare (2015) reporting on the undercount of young children. This calculation is sometimes labeled "net Census coverage error" in other research. In this publication, a negative number consistently implies a net undercount and a positive number implies a net overcount. I chose to use the net Census coverage error construction because I feel having an undercount reflected by a negative number is more intuitive.

When figures are stated in the text as an undercount or an overcount, the positive and negative signs are not used. In converting the difference between Census counts and Demographic Analysis or Dual-Systems Estimates to percentages the difference is divided by the DA or DSE estimate not the Census figure.

Another point of potential confusion is the name applied to the Dual Systems Estimation method. The DSE has been called by different names in the past three Censuses. In the 1990 Census it was called the Post-Enumeration Survey (PES), in the 2000 Census it was called Accuracy and Coverage Evaluation (A.C.E.) and in the 2010 Census it was called Census Coverage Measurement (CCM). In the 2020 Census, this method will again be called the Post-Enumeration Survey or PES (U.S. Census Bureau 2017b). I use the term DSE for consistency.

1.3.1 Net Undercounts, Omissions, and Hard-to-Count Populations

Another term that is related to net undercounts or Census omissions is "hard-to-count" populations. Many closely related terms (hard-to-count areas, hard-to-count populations, difficult to enumerate populations, and hard-to-survey populations) have been used almost interchangeably (Tourangeau et al. 2014). Census Bureau (2017a) also uses the term "Hard-to-Reach" populations to identify groups that are difficult to enumerate accurately.

The U.S. Census Bureau (2017a, p. 2) defines hard-to-count populations as,

> Hard-to-count populations face physical, economic, social, and cultural barriers to participation in the Census and require careful consideration as part of a successful communications strategy.

While this is a good conceptualization of "hard-to-count" populations, it does not specify how to measure the concept and it does not mean that hard-to-count groups necessarily are undercounted in the Census. Generally, groups that have a significant net undercount are thought of as hard-to-count groups, but not all hard-to-count groups have measurable net undercounts in the Census.

Many of the hard-to-count populations not covered by DA and DSE can be addressed to some level by identifying who lives in hard-to-count neighborhoods (O'Hare 2015). In addition, Mail Return Rates are often used as a proxy for Census coverage (Letourneau 2012; Word 1997; Erdman and Bates 2017). A set of hard-to-count factors were provided by Bruce and Robinson (2003) in the mid-1990s that are sometimes used to identify vulnerable populations. Following the 2010 Census the Census Bureau produced a new metric for identifying hard-to-count areas which is called the Low-Response Score (Erdman and Bates 2017).

But it is important to note that hard-to-count factors, Low-Response Scores, and Mail Return Rates (Bruce and Robinson 2003; Erdman and Bates 2017) are not measures of Census coverage per se. Moreover, the association between Mail Return Rates and net undercounts are not always clear. O'Hare (2016, p. 51) shows that only five of thirteen groups concentrated in neighborhoods with low Mail Return Rates had a net undercount rate that was statistically significantly different than zero.

1.4 Perspectives on Differential Undercounts

Differential undercounts suggest comparisons between the coverage rate of one group and the coverage rate of another group. But sometimes it is not clear what the appropriate comparison group should be and sometimes data for the appropriate comparison group are not available. For example, if one is looking at the net undercount rate for young children, should that undercount rate be compared to the net undercount rate for the total population, the net undercount rate for some age-group of adults, or the net undercount rate for the elderly?

Other times, there is no net undercount measurement for the group one would like to use as a comparison group. For example, for the 2010 DA undercount estimates, there are no estimates for the White population or the Non-Hispanic White population. So, demographers often compare Census coverage of the Black population to the Non-Black population. Demographers sometimes end up making comparisons that may not be the most appropriate ones from a conceptual point of view but are dictated by the data available.

Comparing all net undercount rates to the net undercount rate for the total population would provide one constant benchmark, but it would often overlook crucial differences between groups. For example, from a social justice point of view, the difference between the net undercount rate of Blacks and Non-Hispanic Whites is probably more meaningful than the difference between net undercount rates of Blacks and the total population. Keep in mind the total population includes many other hard-to-count groups besides Blacks.

If this publication were only focused on one group, the appropriate comparison group might be easy to identify. But the net undercounts and net overcounts for many groups are compared in this book. Therefore, I focus on identifying groups with high net undercount or omissions rates even if there is not a specific comparison group. One point of this book is giving readers a good sense of which groups have the highest net undercounts and omissions rates in the U.S. Census.

1.5   Contents of This Book

Following this Introductory Chapter, I provide a Chapter on the uses of Census data. It is difficult to understand the importance of Census accuracy unless one understands how Census data are used in the public and private sectors. Census undercounts are important because data accuracy is linked to many equity and social justice issues. In addition, Census undercounts are one of the biggest Census Bureau public relations issues. With respect to Census undercounts Kissam (2017, p. 797) states, "These persistent undercounts raise difficult questions about why it occurs and are troubling due to their practical implications for the conduct of public policy."

In Chap. 3, I describe the key methods used by the Census Bureau to assess accuracy in the Decennial Census. The descriptions offered here are relatively parsimonious with citations provided for readers who would like more detailed information on this topic. The strengths and limitations of each method are also noted.

In Chap. 4, I provide a summary of key undercount differentials based on age, sex, race, Hispanic Origin, and tenure. This Chapter is meant to give readers an overview of Census undercounts and provide a framework and a foundation for several later Chapters. In Chap. 4, information is also presented on some of the key groups not reflected in the Census Bureau's DA or DSE data.

The next several Chapters focus on specific demographic characteristics. I present the groups in the order in which the questions appear in the Census questionnaire.

In Chap. 5, I examine undercount differentials by age. The major points here are the very high net undercount of young children and the net overcounts of young adults and the older population (age 60 plus). In Chap. 6, I explore differences by sex, which shows, on average, males have a net undercount while females have a net overcount.

In Chaps. 7 through 11, I explore differentials in Census accuracy by major race/Hispanic Origin groups (Hispanics, Blacks, Asians, American Indians/Alaskan Native, and Native Hawaiian/Pacific Islander). Researchers differ in their views about the appropriateness of using race and Hispanic Origin as a lens for differential undercounts. Some feel differences by race should not be the focus of attention because racial differences in net undercounts are simply a product of other nonracial factors and we should be focused on these other more dominant non-racial factors. For example, Schwede et al. (2014, p. 293) state,

> Though there is no reason to believe that race or ethnicity in and of itself leads to coverage error, it seems that some underlying variables associated in past studies with undercounting may also be correlated with race (e.g., mobility, complex living situations, and language isolation).

In other words, race is widely seen as a proxy for a combination of factors related to the likelihood of being missed in the Census.

Nonetheless there are least three reasons for looking at differential Census coverage through the lens of race and Hispanic Origin. First, at least from a civil rights perspective, differential undercounts among minority groups is a central problem with the Decennial Census results and many of the most pronounced differentials occur among different race groups. Second, there is a wealth of data related to Census coverage by racial categories. Examination of net undercounts by race is a very prominent facet in much of the previous work on differential undercounts. Third, several analyses show that race is still a salient factor even after many of the other factors are controlled (Erdman and Bates 2017; Fernandez et al. 2018).

Readers may also note that there is no Chapter on the White or Non-Hispanic White population. The White population is seldom discussed in the context of Census undercount because the Whites are almost always counted more accurately than their counterparts in racial and Hispanic minority groups. In this publication, data for the Non-Hispanic White population are provided in many Chapters as a comparison group for minority groups.

In Chap. 12, I examine Census coverage rates by tenure. Homeownership often conveys a status and commitment to place that impacts Census participation. Data consistently show that renters have net undercount rates while homeowners have net overcounts. Examination of differential undercounts by tenure is also a reflection of socioeconomic status since homeowners generally have higher incomes than renters.

In Chap. 13, theories and data on why people are missed in the Census are examined. The material in this Chapter includes broad frameworks for understanding why people are missed in the Census as well as several individual mechanisms that may result in someone being left out of the Census count.

Given the pervasive and long-standing differential undercount patterns, it is worth noting efforts the Census Bureau has made to reduce differential undercounts. In Chap. 14, some of the key activities the Census Bureau has utilized in the past few decades to improve Census coverage and reduce differential undercounts are reviewed.

In Chap. 15, some of the key issues surrounding the upcoming 2020 Census are examined. Some of the major developments regarding 2020 Census planning are discussed and I offer some thoughts about the implications these have for Census coverage in the 2020 Census.

Finally, in Chap. 16 I offer a brief summary of key findings and some of their implications.

Most of the focus in this book is on data from the 2010 Census with some examination of long-term trends when the data are available and appropriate. The reality is many of the groups that have high net undercounts in the 2010 Census have experienced problems with Census coverage for many decades. I feel it is important to recognize that history.

1.6 Summary

While the Census Bureau tries very hard to count every person in the country some people are always missed, and some groups are missed at a higher rate than others. The net undercount reflects the difference between people missed and people counted twice. Some groups have higher net undercount rates than others in the Census and some groups have higher omissions rates than other groups. This book focuses primarily on groups with relatively high net undercounts and high omissions rates and to a lesser extent on groups with relatively high net overcounts.

References

Blank, R. (2012). Statement by Deputy U.S. Commerce Secretary Rebecca Blank on Release of the Data Measuring Census Accuracy. May 22, 2012.

Bruce, A., & Robinson, J. G. (2003). *The planning database: Its development and use as an effective tool in census 2000*. Paper presented at the Annual Meeting of the Southern Demographic Association, Arlington, VA.

de la Puente, M. (1993). *Using ethnography to explain why people are missed or erroneously included by the census: Evidence from small area ethnographic research*. U.S. Census Bureau

Erdman, C., & Bates, N. (2017). The low response score (LRS) a metri to locate, predict, and manage hard-to-survey populaiton. *Public Opinion Quarterly, 81*(1), 144–156.

Fay, R. E, Passel, J. S., Robinson, J. G., & Cowan, C. D. (1988). *The coverage of the population in the 1980 census*. U.S. Census of Population and Housing, Evaluation, and Research Reports, PHC80-E4, U.S Census Bureau, Washington, DC.

Fernandez, L., Shattuck, R., & Noon, J. (2018). The use of administrative records and the American Community Survey to study the characteristics of undercounted young children in the 2010 Census. Center for Administrative Records Research and Applications, CARRA Working Paper Series, Working Paper Series #2018-05, U.S Census Bureau. Washington, DC.

Govern, K., Coombs, J., & Glorioso, R. (2012). *2010 census coverage followup assessment report.* U.S. Census Bureau 2010 Census Program for Evaluation and Experiments, No. 244, March 29.

Jensen, E., Benetsky, M., & Knapp, A. (2018). A sensitivity analysis of the net undercounts for young Hispanic children in the 2010 census. In *Poster at eh 2018 Population Association of American Conference*, Denver, Colorado April 25–28 downloaded May 5, 2108, at https://paa.confex.com/paa/2018/meetingapp.cgi/Paper/20826.

King, H., Ihrke, D., & Jensen, E. (2018). Subnational estimates of net coverage error for the population aged 0 to 4 in the 2010 census. In *Paper present the 2018 Population Association of American Conference*, April 25–28, Denver Colorado, Downloaded May 6, 2018 https://paa.confex.com/paa/2018/meetingapp.cgi/Paper/21374.

Kissam, E. (2017). Differential undercount of Mexican immigrant families in the U.S. census. *Statistical Journal of the International Association of Official Statistics*, 797–816. https://doi.org/10.3233/sji-170366m (IOS Press).

Letourneau, E. (2012). *Mail response/return rates assessment.* 2010 Census Planning Memorandum Series, No. 198, U.S. Census Bureau, Washington, DC.

Mayol-Garcia, Y., & Robinson, G. (2011). Census 2010 counts compared to the 2010 population estimates by demographic characteristics. In *Poster presented at the Southern Demographic Association Conference, October, Tallahassee, FL.*

O'Hare, W. P. (2015). *The undercount of young children in the U.S. Decennial Census.* Springer Publishers.

O'Hare, W. P. (2016). Who lives in hard-to-count neighborhoods? *International Journal of Social Science Studies, 4*(4), 43–55.

Raymondo, J. C. (1992). *Population estimation and projection: Methods for marketing, demographic, and planning personnel.*

Robinson, G, J., & Adlaka, A. (2002). *Comparison of A.C.E. revision II results with demographic analysis,* DSSD A.C.E. Revision II Estimates Memorandum Series #PP-41, December 31, 2002, U.S. Census Bureau, Washington, DC.

Robinson, G. J., Bashir. A., Das Dupta, P., & Woodward, K. A. (1993). Estimates of population coverage in the 1990 United States U.S. Decennial Census based on demographic analysis. *Journal of the American Statistical Association, 88*(423), 1061–1071.

Robinson, J. G. (2010). *Coverage of population in Decennial Census 2000 based on demographic analysis: The history behind the numbers.* Decennial Census Bureau, Working Paper No. 91, available online at http://www.Census.gov/population/www/documentation/twps0091/twps0091.pdf.

Schwede, L. (2003). *Complex households and relationships in the Decennial Census and in ethnographic studies of six race/ethnic groups.* Final Census 2000 Testing and Experimentation Program Report. Available at: http://www.Census.gov/pred/www/rpts/Complex%20Households%20Final%20Report.pdf.

Schwede, L. (2004). Household types and relationships in six race/ethnic groups: Conceptual and methodological issues. In *Proceedings of the American Statistical Association Section on Survey Research Methods* (pp. 4991–4998). Available at: www.amstat.org/sections/srms/proceedings/y2004/Files/Jsm2004–000772.pdf.

Schwede, L. (2006). Who lives here? Complex ethnic households in America. In *Complex Ethnic Households in America.* Rowman and Littlefield.

Schwede, L. (2007). *A new focus: Studying linkages among household structure, race/ethnicity, and geographical levels, with implications for census coverage.* SRD Report RSM 2007/38. Abstract available at: http://www.Census.gov/srd/www/abstract/rsm2007–38.html.

Schwede, L., & Terry, R. (2013). *Comparative ethnographic studies of enumeration methods and coverage across race/ethnic groups.* 2010 Census Program for Evaluations and Experiments Evaluation. 2010 Census Planning Memoranda Series, No. 255. March 29, 2013. http://www.Census.gov/2010Census/pdf/comparative_ethnographic_studies_of_enumeration_methods_and_coverage_across_race_and_ethnic_groups.pdf.

Schwede, L., Terry, R., & Hunter, J. (2014). Ethnographic evaluations on coverage of hard-to-count minority in the US Decennial Censuses. In R. Tourangeau, B. Edwards, T. P. Johnson, K. M. Wolter, & N. Bates (Eds.), *Hard-to-survey populations* (pp. 293–315). Cambridge, MA: Cambridge University Press.

Tourangeau, R., Edwards, B., Johnson, T. P., Wolter, K. M., & Bates, N. (2014). *Hard-to-survey populations.* Cambridge, MA: Cambridge University Press.

U.S. Census Bureau. (1974). *Estimates of coverage of population by sex, race and age: Demographic analysis,* 1970 U.S. Decennial Census of Population and Housing, Evaluation and Research Program, PHC (E)-4.

U.S. Census Bureau. (2012). *2010 components of census coverage for race groups and hispanic origin by age, sex, and tenure in the United States.* DSSD 2010 Census Coverage Measurement Memorandum Series #2010-E-51, U.S. Census Bureau, Washington, DC.

U.S. Census Bureau. (2016). *Developing an integrated communication strategy: Select topics in international census.* U.S. Census Bureau, Washington, DC.

U.S. Census Bureau. (2017a). 2020 census integrated communications plan (Version 1.0) 6/2/2017. U.S. Census Bureau Washington, DC.

U.S. Census Bureau. (2017b). 2020 Census program memorandum series: 2017.17, 2020 Census decision to change the name of the coverage measurement survey to the post-enumeration survey, September 14.

U.S. General Accounting Office. (2003). *2000 census: Coverage measurement programs' results, costs, and lessons learned.* Report GA)-03-287, U.S. General Accounting Office, Washington, DC.

Velkoff, V. (2011). Demographic evaluation of the 2010 census. *Paper presented at the 2011 Population Association of America Annual Conference*, March, Washington, DC.

Williams, J. D. (2012). *The 2010 Decennial Census: Background and issues.* CRS Report for Congress, Congressional Research Services, 7-5700, R4055.

Word, D. L. (1997). *Who responds/ who doesn't? Analyzing variation in mail response rates during the 1990 census.* Population Division Working Paper No. 19, U.S. Census Bureau, Washington, DC.

Open Access This chapter is licensed under the terms of the Creative Commons Attribution 4.0 International License (http://creativecommons.org/licenses/by/4.0/), which permits use, sharing, adaptation, distribution and reproduction in any medium or format, as long as you give appropriate credit to the original author(s) and the source, provide a link to the Creative Commons license and indicate if changes were made.

The images or other third party material in this chapter are included in the chapter's Creative Commons license, unless indicated otherwise in a credit line to the material. If material is not included in the chapter's Creative Commons license and your intended use is not permitted by statutory regulation or exceeds the permitted use, you will need to obtain permission directly from the copyright holder.

Chapter 2
The Importance of Census Accuracy: Uses of Census Data

Abstract This Chapter provides readers with many reasons why the Census count is so important, including the fact that Census data are the backbone of our democratic system of government. In addition, Census-related figures are used to distribute more than $800 billion in federal funding each year to states and localities. Countless decisions in the public and private sectors are based on Census data. Moreover, the impact of flaws in Census counts often last a decade because population estimates, projections, and survey weights, are derived from Census counts.

2.1 Introduction

To understand the importance of differential Census undercounts and omissions it is important to understand how Census data are used. In addition to our scientific and scholarly interest in obtaining correct Decennial Census counts, there are many practical and policy-related reasons why it is important to assess Census coverage. In many cases, Census coverage errors are important because they are both a data problem and a social equity issue.

According to the U.S. Census Bureau (2017), data from the Decennial Census are used for many important applications including:

- Allocating political power
- Distribution of federal funds through funding formulas
- Civil rights enforcement
- Business applications
- Post-Census population estimates and projections
- Providing weights for sample surveys
- Providing denominators for rates
- Community planning
- Economic and social science research.

A more detailed description of how Census data are used is provided in Appendix A of the Census Bureau's 2020 Census Complete Count Committee Guide (U.S. Census Bureau 2018). A number of these points are discussed in more detail below.

© The Author(s) 2019
W. P. O'Hare, *Differential Undercounts in the U.S. Census*,
SpringerBriefs in Population Studies, https://doi.org/10.1007/978-3-030-10973-8_2

2.2   Political Power

Constitutional scholar Leavitt (2018, p. 2) provides a clear idea of the importance of the Census when he states,

> It is impossible to overstate the constitutional significance of the Decennial Census. The requirement that has become a mandate to count each and every individual in the country--- the 'actual Enumeration' of the population in every decade ---is embedded in the sixth sentence of the Constitution. It is the very first act that the Constitution prescribes as an express responsibility of the new federal Government.

The fact that the Decennial Census is mentioned early in the Constitution by the founding fathers, suggests the central role they envisioned for it in our system of governance. Counts from the Census are used to distribute political power both in terms of assigning seats in the U.S. House of Representatives to states based on population and in the judicially mandated one-person/one-vote rule used for constructing political districts (Grofman 1982; McKay 1965; Balinski and Young 1982; Baumle and Poston 2004). The "one person-one vote" rule requires election districts to be equal (or nearly equal) in population size. Calculation of election district population size is almost always based on data from the Decennial Census.

The fundamental relationship between Census counts and political power was summarized by Heer (1968, p. 11) 50 years ago,

> Where a group defined by racial or ethnic terms and concentrated in special political jurisdictions is significantly undercounted in relation other groups. Then individual members of that group are thereby deprived of the constitutional right to equal representation in the House of Representatives and, by inference, in other legislative bodies.

Decennial Census counts for states are used for apportioning the seats in the U.S. House of Representatives and sometimes small differences can be important in determining which state gets the last seat to be assigned (Conk 1987; Baumle and Poston 2018). For example, Crocker (2011) found that if the 2010 Decennial Census count for North Carolina had been 15,753 higher it would have received an additional seat in Congress. Baumle and Poston (2004) also show that small differences in state Census counts can change which state gets the last Congressional seat assigned.

Based on projecting the 2017 Census Bureau population estimates to 2020, Brace (2017) predicts 15 to 17 states will experience changes in their congressional delegation after the post-2020 re-apportionment. However, many states are close to gaining or losing a seat depending on the demographic changes between now and 2020 and the quality of the Census count in each state. Consequently, details about how the Census is conducted could have an impact on apportionment of Congress following the 2020 Census. For example, Baumle and Poston (2018) show that failure to count non-citizens in the 2020 Census would result in several congressional seats changing states in the apportionment following the 2020 Census.

I could not find a definitive number of election districts where Census data are used to draw boundaries for political districts. In addition to the 435 seats in Congress, almost all of the 7383 state legislators are elected from single member districts

(National Conference of State Legislators 2017). Also, nearly ever large city has city council members elected from single-member districts and the same is true for county commissioners. School board members and many special districts also use Census data to construct districts. Therefore, the number of election districts based on Census results must be at least 10,000.

Siegel (2002, Chap. 2) as well as Teitelbaum (2005) provide additional examples of how demographic data are used in a variety of political applications. The bottom line is that any geographic area that is undercounted is not likely to get its fair share of political power (Anderson and Fienberg 2001; Bryant and Dunn 1995).

2.3 Distribution of Public Funds

Decennial Census data are also used in many federal funding formulas that distribute federal funds to states and localities each year (Murray 1992; U.S. Senate 1992; Reamer 2010; Blumerman and Vidal 2009; Hotchkiss and Phelan 2017). Recent research indicates Census-derived data were used to distribute more than $850 billion to states and localities through 302 programs in Fiscal Year 2016 (Reamer 2018). Table 2.1 shows data for the 16 largest federal programs that use Census-derived data to distribute funds. Places that experience a net undercount do not receive their fair share of formula-driven public resources (PriceWaterhouseCoopers 2001). The distribution of federal funds based on Census data will impact some groups more than others. The Annie E. Casey Foundation (2018) shows that these funding formulas are particularly important for programs that support needy children.

Moreover, the amount of federal money given out through funding formulas has increased in recent years. The increase is heavily driven by Medicaid and Medicare Part B where health care costs are rising faster than inflation and the large baby boomer generation is aging and expanding the number of recipients in these programs.

One question that always arises in this area is," How much money does a state lose for each uncounted person?" There is no definitive answer to this question, but an analysis by Reamer (2018) shows that for five programs that use the Federal Matching Assistance Percentage (FMAP) states, on average, would lose $1091 each year for each uncounted person. In some states the figure was lower, and, in some states, it was higher. These figures only apply to the 37 states which are not already at the minimum FMAP value of 50 percent and only for five programs. Another study following the 2000 Census that was focused on 169 metropolitan areas concluded the loss over the 2002–2012 period was $3392 per uncounted person in these jurisdictions but the authors note this estimate is conservative because not all programs are included (PriceWaterhouseCooper 2001, p. ES-1). In another report focused on Idaho (Miller 2018) concludes, "It's estimated each person counted brings about $1200 per year in federal funding to state and local government."

Table 2.1 Largest 16 federal assistance programs that distribute funds on basis of decennial census-derived data, Fiscal Year 2015

Program name	Department	Obligations
Medical Assistance Program (Medicaid)	HHS	$3,11,97,57,66,352
Supplemental Nutrition Assistance Program (SNAP)		$69,48,98,54,016
Medicare Part B (Supplemental Medical Insurance)–Physicians Fee Schedule Services	HHS	$64,17,67,25,988
Highway Planning and Construction	DOT	$38,33,19,04,422
Section 8 Housing Choice Vouchers	HUD	$19,08,75,49,000
Title I Grants to Local Education Agencies (LEAs)	ED	$13,85,91,80,910
National School Lunch Program	USDA	$11,56,08,52,485
Special Education Grants (IDEA)	ED	$11,23,31,12,681
State Children's Health Insurance Program (S-CHIP)	HHS	$11,08,91,52,000
Section 8 Housing Assistance Payments Program (Project-based)	HUD	$9,23,80,92,008
Head Start/Early Head Start	HHS	$8,25,91,30,975
Supplemental Nutrition Program for Women, Infants, and Children (WIC)	USDA	$6,34,76,80,031
Foster Care (Title IV-E)	HHS	$4,63,57,33,000
Health Center Program	HHS	$4,18,14,07,055
Low Income Home Energy Assistance (LIHEAP)	HHS	$3,37,02,28,288
Child Care and Development Fund—Entitlement	HHS	$2,85,86,60,000
Total		$5,89,69,50,29,211

Source Reamer (2017)

2.3.1 Federal Distribution 2015–2030 Based on Census-Derived Figures

Reamer (2018) estimates that in Fiscal Year (FY) 2016 the federal government distributed at least $865 billion to states and localities based on Census-derived data. The $865 billion figure is based on the largest 35 federal programs that use Census-derived data to distribute money and there are many additional programs that are not included in this figure. In 2010 Reamer (2010) estimated that the federal government distributed about $420 billion to states and localities based on Census-derived data in Fiscal Year (FY) 2008. The $420 billion in FY 2008 amounts to $465 billion in 2015 dollars. Consequently, there was an increase from $465 billion to $865 billion between FY2008 and FY2015. Some of the increase from FY2008 to FY2015 is based on adding programs to the calculation and some of the increase is based on increased spending by programs identified in 2008.

The increase between FY2008 and FY2015 amounts to a little more than 11 percent per year. Data shown in Table 2.2 indicate what would happen between

Table 2.2 Hypothetical distribution of federal funds based census-derived data: FY 2015 to FY2030

	In billions of 2015 dollars
FY2008[a]	$465
FY2015[b]	$844
FY2016	$937
FY2017	$1040
FY2018	$1154
FY2019	$1281
FY2020	$1422
FY2021	$1579
FY2022	$1752
FY2023	$1945
FY2024	$2159
FY2025	$2396
FY2026	$2660
FY2027	$2953
FY2028	$3277
FY2029	$3638
FY2030	$4038
Total 2021–2030	$26,398

[a]*Source* Reamer (2010)
[b]*Source* Reamer (2018)

FY2015 and FY2030 if there were an 11 percent increase per year in the amount of federal funds distributed based on Census-derived data. Perhaps most importantly in the decades following the 2020 Census more than $26 trillion dollars could be distributed to states and localities on the basis of 2020 Census-derived data.

Of course, no one knows exactly what will happen in the future regarding the distribution of federal funds based on Census-derived data and the measurement of change between FY2008 and FY2105 is not precise, but the scenario reflected in Table 2.2 provides one plausible trajectory. Moreover, even if the projected dollars in Table 2.2 are too high by 10 or 20%, the amount of money distributed between 2020 and 2030 using Census-derived data is still enormous.

Demographic data are also used to distribute state government funds within states, but I was unable to find a good estimate of how much money is regularly distributed by state governments based on Census data.

2.4 Population Estimates, Projections, and Surveys

The undercounts in the Decennial Census also have implications for post-Census population estimates and projections. The Census Bureau's post-Census population estimates program, which produces yearly national, state, and county population estimates, uses data from the Decennial Census as the starting point to produce post-Census estimates (U.S. Census Bureau 2014a). If an age cohort is undercounted in the Census, that cohort will be under-represented in the Census Bureau's population estimates for the next decade.

Many population projections also start with the Decennial Census counts, so undercounts in the Decennial Census are likely to be reflected in projections for many years (U.S. Census Bureau 2014b). State population projections, such as those available from the University of Virginia's Weldon Cooper Center for Public Service (2013), are also affected by Census undercounts. In discussing where to get data for state and local projections Smith et al. (2001, p. 113) indicate, "The most commonly used source--and the most comprehensive in terms of demographic and geographic detail—is the Decennial Census of population and housing."

Decennial Census results and the Census Bureau's post-Census population estimates are often used to weight sample surveys both inside and outside government. If the Decennial Census counts and subsequent population estimates underestimate a population group, the weighted survey results will reflect this error (Jensen and Hogan 2017; O'Hare and Jensen 2014; O'Hare et al. 2013). Several analysts have shown how Census undercounts distort estimates of poverty rates for children (Hernandez and Denton 2001; Daponte and Wolfson 2003; O'Hare 2017).

In addition, data from the Census Bureau are often used as denominators for constructing rates such as the child mortality rate. Census undercounts may skew such rates. For example, the 2010 U.S. death rate for all children age 1–4 in 2010 was 26.5 per 100,000 and for Hispanic children age 1–4 it was 22.7 per 100,000 (Murphy et al. 2013). These rates are based on using the Census counts as denominators. If one had used the DA estimates for the population age 1–4 instead of the Census counts, the death rate for all children age 1–4 would have been 25.3 (rather than 26.5) and the rate for Hispanic children age 1–4 would have been 20.9 (rather than 22.7). This represents a 5% difference for all children and an 8 percent difference for Hispanic children. This shows how Census undercounts can lead to flawed rates.

2.5 Using Census Data for Planning

Data from the U.S. Decennial Census counts as well as estimates and projections which are based on the Census are used for many planning activities including schools (Edmonston 2001; McKibben 2007, 2012). Flaws in the Census counts can lead to inefficient use of public funds. For example, the high net undercounts of young children in many large cities and urban counties are likely to compromise school planning in those areas (O'Hare 2015).

Census data are also used in health care planning (Koebnick et al. 2012). For example, the Center for Medicare and Medicaid Services (2018) shows how Census are used in health care planning and delivery in rural America.

2.6 Use of Census Data in Business

Census data have been used in business planning as well (Headd 2003). Among other uses, Census data are used by business to determine where to start or expand a business and to determine potential customers for new products. A recent U.S. Department of Commerce publication (2015, p. 2) identifies several business and commercial uses of Census data including;

- Create effective marketing or merchandising strategies to better serve customers and communities
- Inform hiring decisions and workforce evaluation
- Forecast growth and sales to make better strategic decisions
- Stock shelves with the goods suited to local customers preferences
- Invest in infrastructure improvements
- Perform risk analysis.

According to the National Research Council (1995, p. 292);

Retail establishments and restaurants, banks and other financial institutions, media and advertising, insurance companies, utility companies, health care providers, and many other segments of the business world use Census data.

One business group working on the 2020 Census, Council for Strong America (2018), states

A thriving economy relies on timely information about the U.S. population and how it is shifting and changing throughout the country. The Decennial census provides the broadest set of data about residents in the United States that no other body produces.

It is also important to note that many of the data products or data systems used by businesses depend on Census data as a benchmark. In the data-driven and digital-driven work of business decision making, at least one business leader (McDonald 2017, p. 2) recognizes the important role the Census plays.

In such a digital-driven world, the Decennial counting of noses known as the U.S. Census may seem irrelevant or outdated. But in fact, the data that the Census Bureau collects –both in its Decennial count and the annual American Community Survey (ACS) – have never been more important to business constituencies.

2.7   Use of Census Data in Civil Rights Protection

For many groups, the Census is seen a civil rights issue (Leadership Conference on Civil Rights 2017). In addition to heavy use of Decennial Census data in the context of redistricting and voting rights, data from the Census are used to examine equality in jobs and education opportunities. A flawed Census can undermine the ability to examine such issues fairly. According to the Leadership Conference Education Fund (2017, p. 1), "Federal agencies rely on Census and American Community Service (ACS) data to monitor discrimination and implement civil rights laws that protect voting rights, equal employment opportunity, and more."

In addition to the use of Census data for many obvious civil rights purposes, it is also used for some lessen known civil rights programs. For example, under section 203 of the Voting Rights Act, data from the Census Bureau are used to identify jurisdictions that must provide language assistance in voting that is based on the number of people in the jurisdiction that speak a language other than English (Advancing Justice 2016).

2.8   Public Perceptions of Growth or Decline

High net undercounts can provide misleading public impressions about the size or growth of the population in a given location. And these perceptions can have a significant impact on public and private investment decisions related to a community.

This point is difficult to quantify but in many instances the size of a population translates into the importance politicians and marketers give it. In response to the 2000 Census, one public official stated, "Pride in the community is involved. I want people to really know how big we are. We aren't just a little burgh in south Louisiana" (cited in Prewitt 2003, p. 7). If communities are perceived as losing population because of an undercount, it can affect the willingness of investors to put money in those communities.

2.9   Science and Scholarship

West and Fein (1990) as well as Clogg and colleagues (1989) review several ways in which the Decennial Census undercounts affect social science research results. Clogg and his colleagues (1989, p. 559) conclude, "Because undercount rates (or coverage rates) vary by age, race, residence and other factors typically studied in social science research, important conceptual difficulties arise in using Decennial Census results to corroborate sampling frames or to validate survey results."

2.10 Census Planning

Finally, to improve Census-taking procedures in the future, it is important to understand which groups are undercounted at the highest rates in the past Censuses. Information on net undercounts and omissions have been used by the Census Bureau to improve the Census-taking procedure from decade to decade. For example, noting the high net undercount of young children in the 2010 Census prompted the Census Bureau to develop plans to reduce the net undercount of young children in the 2020 Census (Jarmin 2018; Walejko and Konicki 2018).

2.11 Summary

Data from the Decennial Census are used for many important applications including:

- Allocating political power
- Distribution of federal funds through funding formulas
- Population estimates and projections
- Providing weights for sample surveys
- Providing denominators for rates
- Civil rights enforcement
- Public and private sector planning
- Economic and social science research
- Improving the accuracy of the Census over time.

It is clear from the content of this Chapter that the Census is more than just a statistical exercise. Census data are used in some of the most important aspects of our society including our system of governance, distribution of federal dollars for many important programs, and thousands of public and private sector decisions.

References

Advancing Justice. (2016). *Census director identifies jurisdictions that must provide language assistance under Section 203 of the voting rights act*. National Association of Latino Elected Officials, December.

Anderson, M., & Fienberg, S. E. (2001). *Who Counts?*. Russell Sage Foundation, New York: The Politics of Census Taking in Contemporary America.

Balinski, M., & Young, H. P. (1982). *Fair representation: Meeting the ideal of one man, one vote*. New Haven, CT: Yale University Press.

Baumle, A. K., & Poston, D. L. (2004). Apportioning the house of representatives in 2000: The effects of alternative policy scenarios. *Social Science Quarterly, 85*(3), 578–603 (September).

Baumle, A. K., & Poston, D. L. (2018). *Alternative house reapportionment scenarios based on projected 2020 census results*. Paper presented at the Population Association of American Conference, Denver CO., April 27.

Blumerman, L. M., & Vidal, P. M. (2009). *Uses of population and income statistics in federal funds distribution—with a focus on census bureau data*. Government Divisions Report Series, Research Report #2009-1. Washington, DC: U.S. Census Bureau.

Brace, K. (2017) *Some changes in apportionment allocations with new 2017 census estimates: But greater change likely by 2020*. Election Data Services, December 20.

Bryant, B. E., Dunn, W. (1995). *Moving power and money: The politics of U.S. decennial census taking*. Ithaca, NY: New Strategists Publications.

Center for Medicare and Medicaid Services. (2018). *Rural health clinic*. Medicare Learning Network Fact Sheet, January.

Clogg, C. C., Massaglie, M. P., & Eliason, S. R. (1989). Population undercount and social science research. *Social Indicators Research, 21*(6), 559–598.

Conk, M. (1987). *According to their respective numbers*. New Haven, CT: Yale University Press.

Council for Strong America. (2018). Business depends on an accurate census. https://www.strongnation.org/readynation/2020-census.

Crocker, R. (2011). *House apportionment 2010: States gaining, losing and on the margin*. Congressional Research Service, 7-5700 R41584.

Daponte, B. O., & Wolfson, L. J. (2003). *How many American children are poor? Considering census undercounts by comparing census to administrative data* (Unpublished paper).

Edmonston, B. (2001). *Effects of U.S. decennial census undercoverage on analyses of school enrollments: A case study of Portland public schools*. U.S. Census Monitoring Board, Report Series, Report No. 5, February.

Grofman, B. (1982). *Representation and redistricting issues*. Lexington, MA: Lexington Books.

Headd, B. (2003). Redefining business success: Distinguishing between closure and failure. *Small Business Economics, 21*(1), 51–61.

Heer, D. M (Ed.), (1968). *Social statistics and the city*. Cambridge, MA: Joint Center for Urban Studies.

Hernandez, D. & Denton, N. (2001). *Census affects children in poverty*. Washington, DC: U.S. Census Monitoring Board.

Hotchkiss, M., & Phelan, J. (2017). *Uses of census bureau data in federal funds distribution*. Washington, DC: U.S. Census Bureau (September).

Jarmin, R. (2018). *Improving our count of young children*. Census Bureau's Directors Blog July 2. https://www.Census.gov/newsroom/blogs/director/2018/07/improving_our_count.html.

Jensen, E., & Hogan, H. (2017). The coverage of young children in demographic surveys. *Statistical Journal of the International Association of Official Statistics, 33,* 321–333.

Koebnick, C., Langer-Gould, A., Gould, M. K., Chao, C. R., Iyer, R. L., Smith, N., et al. (2012). Sociodemographic characteristics of members or a large integrated health care system: Comparison with US census bureau data. *The Permanente Journal, 16*(3), 37–41.

Leadership Conference Education Fund. (2017). Fact sheet: The census and civil rights. Downloaded on June 13, 2017 from http://civilrightsdocs.info/pdf/Census/Fact-Sheet-Census-and-Civil-Rights.pdf.

Leavitt, J. (2018). *Testimony before the United States house of representatives, committee on oversight and government reform*. Progress Report on the 2020 Census, May 8.

McDonald, S. (2017). A 2020 census flop would pose a danger to U.S. Business, #Bigdata December 6.

McKay, R. (1965). *Reapportionment: The law and politics of equal representation*. New York, NY: The Twentieth Century Fund.

McKibben, J. (2007). *The use of school enrollment data to estimate census undercounts in small areas, presentation and applied demography conference* (p. 2007). TX January: San Antonio.

McKibben, J. (2012). *Using school enrollment data to measure small area coverage rates of the 2010 census, presentation and applied demography conference* (p. 2012). TX January: San Antonio.

Miller, C. (2018). Cited in Idaho Census 2020 Planning underway. *Idaho Business Review* (March 8, 2018), written by Sharon Fisher.

Murphy, S. L., Xu, J., & Kochanek, K. D. (2013). *Deaths: Final data for 2010. National Vital Statistics Reports, 61*(4).

Murray, M. P. (1992). Census adjustment and the distribution of federal spending. *Demography, 29*(3), a319–332,

National Conference of State Legislators. (2017). Downloaded June 11, 2017 from http://www.ncsl.org/research/about-state-legislatures/number-of-legislators-and-length-of-terms.aspx.

National Research Council. (1995). *Modernizing the U.S. Census.* Washington, DC: National Academy Press.

O'Hare, W. P. (2015). *The undercount of young children in the U.S. decennial census.* Springer.

O'Hare, W. P. (2017). *The impact of the undercount of young children in the census on poverty estimates from the American community survey.* Presentation at American Community Survey Users Conference, 2017, Alexandria, VA.

O'Hare, W. P., & Jensen, E. B. (2014). *The representation of young children in the American community survey.* Presentation at the ACS Users Group Conference, Washington, DC, May 29–30.

O'Hare, W. P., Jensen, E., & O'Hare, B. C. (2013). *Assessing the undercount of young children in the U.S. decennial census: Implications for survey research and potential explanations.* Paper presented at the 2013 American Association of Public Opinion Researchers Annual Conference, Boston, MA.

Prewitt, K. (2003). Politics and science in census taking. In series *The American People: Census 2000.* Russell Sage Foundation and Population Reference Bureau, Population Reference Bureau, Washington, DC.

PriceWaterhouseCooper. (2001). *Effect of U.S. Decennial Census2000 Undercount on Federal Funding to States and Selected Counties, 2001–2012.* Report to the U.S. Census Monitoring Board, Presidential Members.

Reamer, A. D. (2010). *Counting for dollars: The role of the decennial census in the geographic distribution of federal funds.* Washington, DC: Brookings Institution, Metropolitan Policy Program.

Reamer, A. D. (2017). *Counting for dollars.* Washington, DC: George Washington University.

Reamer, A. D. (2018). *Census-guided federal assistance to rural America, Report #3, Counting for Dollars 2020: The Role of the decennial census in the geographic distribution of federal funds.* Washington, DC: George Washington University (August 24).

Seigel, J. (2002). *Applied demography: Applications to business, government, law, and public policy.* Academic Press.

Smith, S., Tayman, J., & Swanson, D. A. (2001). State and local population projections: Methods and analysis. *The Plenum Series on Demographic Methods and Population Analysis,* Kluwer.

Teitelbaum, M. S. (2005). Political demography. In M. Michlin & D. L. Poston (Eds.), *Handbook of population* (pp. 719–730). Springer.

The Annie E. Casey Foundation. (2018). *2018 Kids count data book: State trends in child well-being.* Baltimore, MD: The Annie E. Casey Foundation.

U.S. Senate. (1992). *Dividing dollars: Issues in adjusting decennial counts and intercensal estimates for funds distribution.* Report prepared by the Subcommittee on Government Information and Regulation of the Committee on Government Affairs, 102nd Congress, 2nd session Senate Print 102-83, U.S. Government Printing Office, Washington, DC.

U.S. Census Bureau. (2014a). (BILL UPDATE). The 2013 population estimates are available online at http://factfinder2.Census.gov/faces/tableservices/jsf/pages/productview.xhtml?src=bkmk.

U.S. Bureau of the Census. (2014b). U.S. Population Projections: 2014–2060, Release Number CB14-TPS.86.

U.S. Census Bureau. (2017). *2020 census operational plan: A new design for the 21st Century, Version 3.0.* Washington, DC: U.S. Census Bureau.

U.S. Census Bureau. (2018). *2020 census complete count committee guide, D-1280.* Washington, DC: U.S. Census Bureau.

U.S. Department of Commerce. (2015). *The value of the american community survey: Smart government, competitive business, and informed citizens, economic and statistics administration.* Washington, DC: U.S. Department of Commerce.

Walejko, G., & Konicki, S. (2018). *Census efforts to reduce the undercount of young children.* Vancouver Canada: Poster Presented at Joint Statistical Meeting.

Weldon Cooper Center for Public Service. (2013). Projections for the 50 states and D.C. Available online at http://www.coopercenter.org/demographics/national-population-projections.

West, K. K., & Fein, D. J. (1990, May). U.S. decennial census undercount: An historical and contemporary sociological issues. *Sociological Inquiry, 60*(2), 127–141.

Open Access This chapter is licensed under the terms of the Creative Commons Attribution 4.0 International License (http://creativecommons.org/licenses/by/4.0/), which permits use, sharing, adaptation, distribution and reproduction in any medium or format, as long as you give appropriate credit to the original author(s) and the source, provide a link to the Creative Commons license and indicate if changes were made.

The images or other third party material in this chapter are included in the chapter's Creative Commons license, unless indicated otherwise in a credit line to the material. If material is not included in the chapter's Creative Commons license and your intended use is not permitted by statutory regulation or exceeds the permitted use, you will need to obtain permission directly from the copyright holder.

Chapter 3
Methodology Used to Measure Census Coverage

Abstract The two primary methods used to assess Census coverage in the U.S. are Demographic Analysis (DA) and Dual-Systems Estimates (DSE). These two methods are introduced in this Chapter along with some of their strengths and weaknesses.

3.1 Introduction

How do we know who is missed in a Census or which groups have a net undercount? Several methods have been used over time and in various countries to answer this question but in the U.S. only the Demographic Analysis (DA) method and the Dual-Systems Estimates (DSE) method provide quantitative answers to the question posed above (Mulry 2014; Hogan et al. 2013; Bryan 2004; Anderson 2004).

According to the U.S. Census Bureau (2012d, p. 2),

> The Census Bureau has historically relied on two principal methods to provide measures of the quality of each Census. One method is based on a post-enumeration survey, which is the topic of this report. The other method is based on demographic analysis, which uses various types of demographic data in order to build an historical account of population change.

Briefly, DA compares the Census count to an independent estimate of the expected population based on births, deaths, and net international migration. The DSE method uses a Post-Enumeration Survey to independently gather information on people that can be compared to the Census count to assess correct enumerations, omissions, and erroneous inclusions (mostly people counted more than once). Each of these methods is described in the next two sections of this Chapter along with some of their strengths and limitations.

One important difference between DA and DSE data is the level of age/sex detail available. Detailed 2010 data from the DA estimates are available so researchers can construct tables for whatever age-sex-race/Hispanic groups as they wish, within the limits of the data. For the 2010 DA data, one must download files from the Census Bureau and construct their own net undercount/overcount tables.

On the other hand, data from the DSE method are only provided for a few age/sex groups determined by the Census Bureau. The DSE data are provided in a series of reports that provide net undercounts and omissions.

© The Author(s) 2019
W. P. O'Hare, *Differential Undercounts in the U.S. Census*,
SpringerBriefs in Population Studies, https://doi.org/10.1007/978-3-030-10973-8_3

3.2   Demographic Analysis Methodology

Demographic Analysis has been used since the 1950 Census to provide estimates of net undercounts in the U.S. Census. As stated above, this method creates a separate independent estimate of the expected population based on births, deaths, and net international migration and the expected population is then compared to the Census count to determine net undercounts and net overcounts. DA estimates are provided for both males and females, Black and Non-Black, by single year of age. Data on Hispanics are provided for those below age 20 in the 2010 DA estimates.

DA is an example of the cohort-component method of population estimation meaning each component of population change (births, deaths, and migration) is estimated for each birth cohort. The cohort-component method is one of the most widely used techniques in population estimation (United Nations 1970; Bryan 2004). Since there are already several detailed descriptions of the DA methodology available, I will only review the method briefly here (Robinson 2010; Himes and Clogg 1992; U.S. Census Bureau 2010).

The DA method has been used to assess the accuracy of Decennial Census figures for more than a half century (Coale 1955; Coale and Zelnick 1963; Coale and Rives 1973; Siegel and Zelnik 1966). Its origins are often traced back to an article by Price (1947), which found an unexpectedly high number of young men who turned up at the first compulsory selective service registration on October 6, 1940 and alerted demographers to the possibility of under-enumeration in the 1940 Decennial Census.

The DA method employed for the 2010 Decennial Census used one technique to estimate the population under age 75 and another method based on Medicare enrollment to estimate the population age 75 and older (West 2012). The 2010 DA estimates for the population age 0–74 are based on the compilation of historical estimates of the components of population change: Births (B), Deaths (D), and Net International Migration (NIM). The data and methodology for each of these components is described in separate background documents prepared for the development and release of the Census Bureau's 2010 DA estimates (Robinson 2010; Devine et al. 2010; Bhaskar et al. 2010).

As described by the U.S. Census Bureau (2010) the DA population estimates for age 0–74 are derived from the basic demographic accounting Eq. (3.1) applied to each birth cohort:

$$P_{0-74} = B - D + NIM \tag{3.1}$$

P_{0-74} = population for each single year of age from 0 to 74 (people less than a year old are labeled age 0 by the Census Bureau)
B = number of births for each age cohort
D = number of deaths for each age cohort since birth
NIM = Net International Migration for each age cohort.

For example, the estimate for the population age 17 on the April 1, 2010 Decennial Census date is based on births from April 1992 through March 1993, reduced by the deaths to that cohort in each year between 1992 and 2010, and incremented by Net International Migration (NIM) of the cohort each year over the 17-year period.

The birth and death data used in the Census Bureau's DA estimates come from the U.S. National Center on Health Statistics (NCHS) and these records are widely viewed as being accurate and complete. The National Center for Health Statistics (2014, p. 2) states, "A chief advantage of birth certificate data is that information is collected for essentially every birth occurring the country each year…" After a thorough review of vital statistics prior to the 2010 Census, the U.S. Census Bureau (Devine et al. 2010, p. 3) stated:

> The following assumptions are made regarding the use of vital statistics for DA:
>
> - Birth registration has been 100% complete since 1985.
> - Infant deaths were underregistered at one-half the rate of the underregistration of births up to and including 1959.
> - The registration of deaths for ages 1 and over has been 100% complete for the entire DA time series starting in 1935.

In addition to regularly published totals, the Census Bureau receives microdata files from NCHS containing detailed monthly data on each birth and death. These files were used primarily for DA estimates by race. Construction of DA estimates by race is discussed later in this Chapter.

The Census Bureau changed the way it calculated Net International Migration (NIM) for the 2010 set of DA estimates (Bhaskar et al. 2010). The current method relies heavily on data from the Census Bureau's American Community Survey (ACS) where the location of the Residence One Year Ago (ROYA) is ascertained for everyone in the survey age 1 or older. The total number of yearly immigrants is derived from this question in each year of the ACS, and then that total number of immigrants is distributed to demographic cells (sex, age, and race) based on an accumulation of the same data over the last five years of the ACS. Five years of ACS data are used to provide more stable and reliable estimates for small demographic groups. On the other hand, it is important to note the five-year average may mask changes in trends over time. Given changing economic conditions, it would not be surprising if the immigration pattern in the 2008–2010 period differed from the pattern before 2008, however, I suspect such errors would be small.

Statistics on emigration of the foreign-born population from the U.S. are based on a residual method comparing data from the 2000 Decennial Census to later American Community Survey estimates to develop rates and then applying those rates to observed populations (Demographic Analysis Research Team 2010).

Emigration of U.S. citizens (net native migration) is derived by examining Census data from several other countries (Schachter 2008). This method of estimating out migration of the native-born population is problematic for a couple of reasons. Data are not available for every country, and the quality of some foreign censuses is suspect. See Jensen (2012) for more details on measuring net international migration. In 2018, the Census Bureau staff presented a paper with revised data for net native migration of young children based on data from the 2010 Mexican Census (Jensen et al. 2018).

In preparing for the December 2010 DA release the Census Bureau developed five estimation series with differing assumptions about births, deaths, and net international immigration to reflect the degree of uncertainty in the estimates. The estimates from the five series presented in December 2010 range from 305,684,000 to 312,713,000. The middle series of the DA estimates was nearly a perfect match to the 2010 Census count so when the DA estimates were updated in May 2012, only the middle series was updated.

3.3 Dual-Systems Estimates Methodology

The other major source of data on net undercounts and overcounts in the U.S. Decennial Census is the Census Bureau's Dual-Systems Estimates (DSE) method. This section describes the estimation method used in generating the net coverage for the household population from the DSE approach. The DSE method also provides estimates for the other components of Census coverage shown below. According to Hogan (1993) overall Census coverage can be separated in to the four components below;

(1) Erroneous enumerations due to duplication,
(2) Erroneous enumerations (fictitious, out-of-scope, died before Census day, born after Census day),
(3) Whole-person imputations, and
(4) Omissions.

The Dual-System Estimates (DSE) method compares Census results to the results of a Post-Enumeration Survey (PES) which is conducted right after Census data collection has been completed to determine the number and characteristics of people who are omitted or included erroneously (mostly those double-counted).

Nomenclature can be confusing in this arena. The terms Dual-Systems Estimates (DSE) and Post-Enumeration Survey (PES) are often used interchangeably. Moreover, the DSE/PES approach has been given a different name in each of the past three U.S. Censuses. In 2010, it was called Census Coverage Measurement (CCM), in the 2000 Census it was called Accuracy and Coverage Evaluation (A.C.E.) and in the 1990 Census it was called the Post-Enumeration Survey (PES). Sometimes the DSE or PES approach is simply called the "survey method." The DSE operation in the 2020 Census will be called PES again (U.S. Census Bureau 2017).

There is a long history of using Dual-System Estimation in measuring coverage errors in a Census (Hogan 1993; U.S. Census Bureau 2004; Wolter 1986). But it is widely believed that DSE estimates that are consistent over time began in 1990. For a detailed explanation of the CCM estimation methodology used in the 2010 Census, see Mule (2008).

Dual-System Estimation is based on what is sometimes referred to as a capture-recapture methodology. The Census is the first system or first capture point and the Post-Enumeration Survey is the second capture point. To estimate the number of people correctly included in the Census, one must take a sample from Census enumerations to match to the PES. In the 2010 Census the sample from the Census is referred to as the Enumeration or E-sample and the Post-Enumeration Survey is used to make the second capture and the population in the Post-Enumeration Survey is referred to as the Population or P-sample.

The 2010 CCM program involved a complex sample of about 170,000 housing units in a sample of Census blocks nationwide (Mule 2010). In every sampled block, Census staff did an independent listing of housing units and independent roster of every person living in those housing units as of April 1, 2010, which were then compared to Census records. Because the DSE figures are based on a sample, sampling error was calculated for each estimate to determine statistical significance. Sampling error is not a major issue for large national groups but for smaller groups and small areas, the sampling errors are often large.

The PES interview is used to determine if the person enumerated in the Post-Enumeration Survey should have been counted in a housing unit on Census day (April 1). By comparing the PES results to the Census, CCM can estimate the number of correct enumerations in the Census. Matching also produces an estimate of the erroneous enumerations. Whole-person imputations are taken from census records.

3.4 Strengths and Limitations of DA and DSE Methods

Both the DA and DSE methods for evaluating Census results have strengths and limitations which are discussed below.

There are four major limitations to DA. First, coverage estimates from DA are routinely only available for the nation as a whole. Because many people move after they are born, estimating coverage for subnational geographic units is difficult. DA only tracks in and out migration at the national level.

The population age 0–9, is an exception to this rule. Subnational analysis can be done for the population age 0–9, because the Census Bureau's population estimates for age 0–9 are not linked to the previous Decennial Census (O'Hare 2014; Mayol-Garcia and Robinson 2011; Robinson et al. 1993; Adlakha et al. 2003, U.S. Census Bureau 2014; King et al. 2018). The 2010 estimates for the population age 0–9, are based on a DA-like method that uses births, deaths, and migration to estimate state and county populations.

Second, DA estimates are only available for a few race/ethnic groups. Historically the estimates have only been available for Black and Non-Black groups. This restriction is due to the lack of race specificity and consistency for data collected on the birth and death certificates historically. The only group that has been identified relatively consistently over time is the Black population, and the residual group is labeled Non-Black. In the 2010 DA program, estimates were produced for Black Alone and for Black Alone or in Combination, but only for the population under age 30.

However, in the past few decades comparisons of Black and Non-Black groups have become more problematic because Hispanics are mostly included in the Non-Black group. The Hispanic population is growing rapidly, and Hispanics have high net undercount rates in the Census.

The 2010 DA estimates also include data for Hispanics for the first time, but only for the population under age 20. Hispanics under age 20 were included in the DA estimates in 2010 because Hispanics have been consistently identified in birth and death certificates since 1990.

The third limitation of the DA estimates is that they only supply net under-count/overcount figures. A net undercount of zero could be the result of no one being missed (omissions) or double counted (erroneous enumerations) or for example, it could be the result of ten percent of the population being missed and ten percent counted twice.

The fourth limitation of the DA methodology is the lack of any measures of uncertainty for the estimates similar to standard errors associated with estimates based on sample surveys. However, it should be noted that in the December 2010 DA release, the Census Bureau released five different estimate series based on five sets of assumptions about births, deaths, and net international migration to reflect some of the uncertainty regarding the DA estimates.

Despite these limitations, DA has been used for many decades, the underlying data and methodology are simple and robust, and it has provided useful information for those trying to understand the strengths and weaknesses of the U.S. Decennial Census. According to Robinson (2000, p. 1), "The national DA estimates have become the accepted benchmark for tracking historical trends in net Census undercounts and for assessing coverage differences by age, sex, and race (Black, all other)."

There are several important limitations of the DSE method which should also be acknowledged. First, is the issue of correlation bias. Correlation bias means the kinds of people who are undercounted in the Census are also likely to be undercounted in the PES. This violates the independence assumption of the DSE methodology. If a group of people are likely to be missed in both the Census and the PES, the undercount estimate for that group will be biased downward. According to Martin (2007, p. 436), "The same groups that are affected by coverage errors in the Census also are affected in demographic surveys conducted by the U.S. Census Bureau and other organizations."

The issue of correlation bias in the DSE approach has been discussed by other researchers (Wolter 1986; Wachter and Freedman 1999: Shores 2002; Shores and Sands 2003; The National Research Council 2009). In the 2010 DSE estimates, the Census Bureau (U.S. Census Bureau 2012c) made adjustment for correlation bias for some groups (Black men) but not for others (Hispanic men or young children).

Second, to tell if an individual was counted correctly in the Census, individuals in the Post-Enumeration Survey must be matched to those in the Census records. This raises a couple of potential problems. First, people don't always provide their names consistently. For example, a person might be listed as John Jones in the Census, and Johnathan Jones Jr. in the PES. Deciding if these two entries are a match is not always clear. Often, but not always, the Census Bureau has additional information like address and birthdates to help with matching. Nonetheless the matching procedure allows potential error.

The DSE approach is also hindered sometimes because there is little or no information for some people included in the Census. For example, as a last resort an enumerator may contact a neighbor to find out about people living in a household and the neighbor may say there are two adults and a young child, but no names or ages are provided. These people are included in the Census, but there is not enough information about them in the Census records to allow matching to the PES records.

Third, the method relies heavily on the memory of individuals. In August or September after the April 1 Census, respondents were asked to list all members of their household as of April 1. For some households and some individuals this is a challenge. As Martin (2007, p. 429) states, "Respondents interviewed months after April 1 may find it difficult to recall accurately when a move occurred."

On the other hand, the DSE method has several advantages. One advantage of the DSE method is that the Census Bureau controls all the data collection (unlike DA where they depend on vital records data). Therefore, the concepts and questions used in the PES can be made identical to those used in the Census. For example, questions about race can be asked the same way in the Census and the PES.

A major advantage of the DSE method is that it can be used to ascertain components of Census coverage such as omissions, and erroneous inclusions. This provides a much richer picture of Census coverage than simply looking at net undercounts and net overcounts.

Because the DSE is based on a carefully drawn sample, the coverage estimates include standard measures of uncertainty. DSE data can provide subnational estimates, although the extent to which this is feasible depends to some extent on sample size. In the 2010 CCM, no state had a net undercount that was statistically significantly different from zero (U.S. Census Bureau 2012b). This is likely related, at least in part, to a relatively small sample is some states.

3.5 Consistencies and Inconsistencies Between DA and DSE Results

Table 3.1 shows differences between net undercount estimates of DA and DSE in the 2010 Census for several age-sex groups. For the most part, the results of DA and DSE are relatively consistent. Generally, the groups that have a high net undercount in DA also have a high net undercount in DSE.

For all adult age groups examined, the differences are less than 1.6 percentage points. However, for the population age 0–4, the difference is 3.9 percentage points and for age 5–9 the difference is 2.5 percentage points.

O'Hare et al. (2016) provide detailed documentation of the consistencies and inconsistencies between DSE and DA estimates for young children and after close examination of the differences in the net undercount estimate for children, O'Hare

Table 3.1 Comparison of 2010 census net coverage error from DA and DSE for demographic groups

	DA net error rates[a]	DSE net error rates[b]	Difference (DA − DSE)
Age 0–4	−4.6	−0.7	−3.9
Age 5–9	−2.2	0.3	−2.5
Age 10–17	0.5	1.0	−0.5
Males ages 18–29	0.3	−1.2	1.5
Females age 18–29	1.5	0.3	1.2
Males age 30–49	−2.3	−3.6	1.3
Females age 30–49	1.7	0.4	1.3
Males age 50+	0.5	0.3	0.2
Females age 50+	2.4	2.4	0.0

[a] *Source* Calculated from U.S. Census Bureau May 2012 Demographic Analysis Release
[b] *Source* U.S. Census Bureau (2012b, Table 13)
A negative sign implies a net undercount. The signs here are reversed from the original publication in order to keep directionality consistent within this publication

et al. (2016, p. 702) conclude, "…the DSE approach may underestimate the net undercount of young children due to correlation bias." The inconsistency of undercount estimates for young children from DA and DSE has been noted before (U.S. Census Bureau 2003, p. v; National Research Council report 2004, p. 254). Most experts agree that DA is a better method for measuring the net undercount of young children (U.S. Census Bureau 2014). Given the problems of the DSE estimates for young children they are not included in the DSE tables in the remainder of this book. To the best of my knowledge there are no plans to change the DSE methodology in the 2020 Census to eliminate this problem.

Based on a special analysis which takes advantage of the strengths of both the DA and DSE methods the U.S. Census Bureau (2016) produced reliable omissions rates for young children and used the same method to produce omissions rates for other age/sex groups. Net undercount rates and omissions rates for several groups defined by age and sex based on adjusted omissions rates are shown in Table 3.2.

Groups that have high net undercount rates typically have relatively high omissions rates, but not always. Note young children (age 0–4) had the highest net undercount rate and the highest omissions rate but males age 18–29 had a relatively low net undercount rate and a relatively high omissions rate. For people in this group, the high omissions rate is balanced by the high erroneous enumeration rate leading to a low net coverage rate. This issue is explored in Chap. 5.

Table 3.2 Comparison of demographic analysis-based estimates of omissions and net coverage error in the 2010 census by age and sex

	Net error rate[a]	Omission rate
Age 0–4	−4.6	10.3
Age 5–9	−2.2	7.3
Age 10–17	0.5	4.8
Males ages 18–29	0.3	7.9
Females age 18–29	1.4	6.4
Males age 30–49	−2.3	7.3
Females age 30–49	1.7	2.9
Males age 50+	0.5	4.0
Females age 50+	2.5	1.9

Source U.S. Census Bureau (2016, Fig. 2)

[a]A negative sign reflects a net undercount. The signs here are reversed from the source report in order to keep directionality consistent within this publication

3.6 Measuring the Net Undercount by Race

Historically, Black is the only race group that has been coded relatively consistently in birth and death certificate data, so it is the only group for which DA estimates could be produced. The residual category is labeled Non-Black.

Different data from the Census have been used to compare Census and DA results for Blacks over time. Prior to the 1980 Census, the U.S. Decennial Census figures that were used to compare with the DA estimates for Blacks were the reported U.S. Decennial Census figures for Blacks. In 1980, the Census Bureau compared the DA estimates to a modified file which assigned people in the "some other race" category to a Black or Non-Black category (Fay et al. 1988). In 1990, the Census Bureau used the race of father from the birth certificate to assign race to newborns and then compared DA estimates for Blacks to the MARS (Modified Age, Race, and Sex) file from the Census. For 2000, the Census Bureau used race of father from the birth certificate to assign race to newborns and then DA estimates were compared to an average of Black alone and Black alone or in combination based on the Census Bureau's modified race file (U.S. Bureau of the Census 2003). While there are some inconsistencies in the way race has been measured from one Census to another, it is generally felt the DA estimates can be used to compare undercount estimates for Blacks since 1950 (National Research Council 2004; Velkoff 2011).

The revised DA estimates issued in May 2012 are used for most of the 2010 DA estimates in this book. In May 2012 the Census Bureau issued revised Demographic Analysis estimates, for the total population, the Black alone population, the Black alone or in Combination population, the Not Black Alone population and the Not Black Alone or in combination population (U.S. Census Bureau, 2012a). The estimates for the Black Alone or in Combination populations were only provided for the population below age 30.

The assessment of Black and Non-Black undercount differentials has always had some methodological issues, but those issues increased dramatically since the 2000 Census when respondents were allowed to mark more than one race. See Robinson (2010) for a good general discussion of issues associated with racial U.S. Office of Management and Budget (1997) classifications in the U.S. Decennial Census and the vital events registers.

Key to the DA method for Blacks and Non-Blacks is making the race categories in the vital events data and the Decennial Census data consistent and there are multiple problems in trying to make data collected in the U.S. Decennial Census racial categories comparable to the race data collected on birth and death certificates. In discussing the use of vital statistics for DA estimates by race the U.S. Census Bureau (Devine et al. 2010, p. 4) conclude, "…developing the estimates for DA race categories comes with a more complex, and substantial set of challenges." For a more detailed description of these problems see O'Hare (2015, Sect. 2.5).

For example, the "Some Other Race" category is a response category for the race question in the U.S. Decennial Census but not in birth or death certificates. So, to make comparisons, people who were in the "some other race" category in the Census had to be re-assigned to one or more of the five major race categories.

A second issue is the fact that U.S. Decennial Census respondents in 2000 and 2010 could mark more than one race but it wasn't until 2003 that the federal government issued new standard birth certificate forms allowing parents to mark more than one race. However, birth and death certificate data are collected by states and states did not all adopt the new forms immediately. DA analysis required that the mixed-race data from the birth (and death) certificates be put into Black and Non-Black categories, based on both single-race and multiple-race reported by mother and fathers

Another issue is that birth certificate forms only record the race of the mother and father while the race of a child is asked directly in the Decennial Census. Thus, for birth certificate data, the race of the newborn must be inferred from the race of the parent(s). This is further complicated by a significant level of missing data. While data on the race of mother is relatively complete, many birth certificates are missing data on the race of the father. In 2009, 19% of birth certificate forms did not contain the race of the father (Martin et al. 2011).

Figure 3.1 shows inconsistencies between Black Alone and Black Alone or in Combination are biggest for the youngest age groups. The youngest age groups are the ones that are dependent on matching the race data from the new birth certificates to the race data from the Census.

Given the issues described above, one should view DA estimates for Blacks (alone or alone or in combination) cautiously. Small differences or small changes over time could be due to methodological issues rather than real changes or differences.

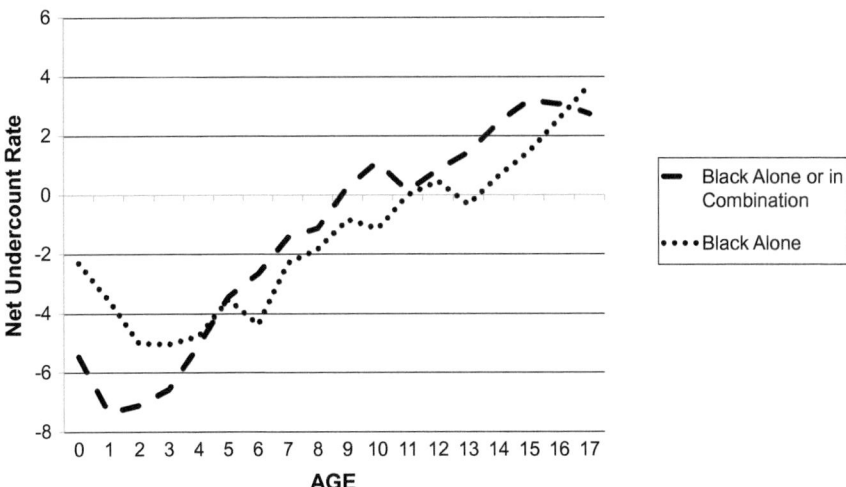

Fig. 3.1 2010 census undercounts for black alone and black alone or in combination by single year of age: 0–17. *Source* U.S. Census Bureau, May 2012 DA Release

3.7 Summary

The main methods for measuring Census undercounts are Demographic Analysis (DA) and Dual-Systems Estimates (DSE). The DA method and the DSE method both have strengths and weaknesses. These two methods produce results that are fairly consistent for all age groups except young children. For the population age 0–4, the DA method estimates a net undercount of 4.6% compared to 0.7% for the DSE method (the DSE method is called Census Coverage Measurement in the 2010 Census). The DA method is widely viewed as the better method for estimates net undercount for young children because it relies heavily on vital events data which are very high quality.

Given the changing methods for identifying the race of people in the Census and in birth/death certificates, and the complications of trying to make the racial categories from the birth certificates consistent with those offered in the Census, DA net undercount estimates for Blacks should be used cautiously.

References

Adlakha, A. L, Robinson, J. G., West, K. K., & Bruce, A. (2003*). Assessment of consistency of decennial census data with demographic benchmarks at the subnational level*. Decennial Census 2000 Evaluation 0.20 Final Report.

Anderson, B. A. (2004). *Undercount in China's 2000 decennial census in comparative perspective.* PSC Research Report, No. 04-565, Population Studies Center, University of Michigan, Ann Arbor, MI.

Bhaskar, R., Scopilliti, M., Hollman, F., & Armstrong, D. (2010). Plans for producing estimates of net international migration for the 2010 demographic analysis estimates. Census Bureau Working Paper No. 90. Available online at http://www.Census.gov/population/www/documentation/twps0090/twps0090.pdf.

Bryan, T. (2004). Population estimates. In J. Siegel & D. Swanson (Eds.), *The methods and materials of demography* (2nd ed., pp. 523–560). Elsevier Academic Press.

Coale, A. J. (1955). The population of the United States in 1950 classified by age, sex and color—A revision of decennial census figures. *Journal of the American Statistical Association, 50,* 16–54.

Coale, A. J., & Rives, N. W. (1973). A statistical reconstruction of black population of the United States: 1880 to 1970: Estimates of true numbers by age and sex, birth rates, and total fertility. *Population Index, 39*(1), 3–36.

Coale, A. J., & Zelnick, M. (1963). *New estimates of fertility and population in the United States.* Princeton, NJ: Princeton University Press.

Demographic Analysis Research Team. (2010). Estimates of net international migration in demographic analysis. *Population Division, U.S. Census Bureau, presentation at 2010 Demographic Analysis Conference, Washington DC, December 6.*

Devine, J., Sink, L., DeSalvo, B., & Cortes, R. (2010). The use of vital statistics in the 2010 demographic analysis estimates. Census Bureau Working Paper No. 88. Available online at http://www.Census.gov/population/www/documentation/twps0088/twps0088.pdf.

Fay, R. E., Passel, J. S., & Robinson, J. G., with Assistance from Cowan, C. D. (1988). *The coverage of the population in the 1980 census.* U.S. Census of Population and Housing, Evaluation and Research Reports, PHC80-E4. Washington, DC: U.S. Census Bureau.

Himes, C. L., & Clogg, C. C. (1992). An overview of demographic analysis as a method for evaluating decennial census coverage in the United States. *Population Index, 58*(4), 587–607.

Hogan, H. (1993). The post enumeration survey: Operations and results. *Journal of the American Statistical Association, 88,* 1047–1060.

Hogan, H., Cantwell, P., Devine, J., Mule, V. T., & Velkoff, V. (2013). Quality and the 2010 census. *Population Research and Public Policy, 32,* 637–662.

Jensen, E. (2012). *International migration and age-specific sex ratios in the 2010 demographic analysis.* Paper presented at the Applied Demography Conference at the University of Texas at San Antonio Texas, January.

Jensen, E., Benetsky, M., & Knapp, A. (2018). *A sensitivity analysis of the net undercount of young hispanic children in the 2010 census.* Poster presented at the Population Association of America annual conference, Denver, CO, April 26–28.

King, H., Ihrke, E., & Jensen, E. (2018). *Subnational estimates of net coverage error for the population aged 0 to 4 in the 2010 census.* Paper delivered at the Population Association of American annual conference, Denver, CO, April 26–28.

Martin, E. (2007). Strength of attachment: Survey coverage of people with tenuous ties to residences. *Demography, 44*(2), 427–440.

Martin, J. A., Hamilton, B. E., & Ventura, S. J. (2011). *Births: Final data for 2009* (Vol. 60, No. 1). National Vital Statistics System; Hyattsville, MD: National Center for Health Statistics.

Mayol-Garcia, Y., & Robinson, J. G. (2011). *Census 2010 counts compared to the 2010 population estimates by demographic characteristics.* Poster presented at the Southern Demographic Association Conference, October, Tallahassee, FL.

Mule, V. T. (2008). 2010 census coverage measurement estimation methodology. DSSD 2010 Census Coverage Measurement Memograndum Series #2010-e-18. Washington DC: U.S. Census Bureau.

Mule, V. T. (2010). *U. S. coverage measurement survey plans.* Paper delivered at the Joint Statistical Meetings, Vancouver Canada.

Mulry, M. (2014). Measuring undercounts for hard-to-reach groups. In R. Tourangeau, B. Edwards, T. P. Johnson, K. M. Wolter, & N. Bates (Eds.), *Hard-to-survey populations*. Cambridge University Press.

National Center for Health Statistics. (2014). *Assessing the quality of medical and health data from the 2003 birth certificate revision: Results from two states* (Vol. 62, No. 2). National Vital Statistics Reports, U.S. Department of Health and Human Services, Centers for Disease Control and Prevention.

National Research Council. (2004). The 2000 census: Counting under adversity. In C. F. Citro, D. L. Cork, & J. L. Norwood (Eds.), *Panel to review the 2000 census* (p. 254). Committee on National Statistics, Division of Behavioral and Social Science and Education. Washington, DC: The National Academy Press.

National Research Council. (2009). Coverage measurement in the 2010 census. In R. M. Bell & M. I. Cohen (Eds.), *Panel on correlation bias and coverage measurement in the 2010 decennial census*. Committee on National Statistics, Division of Behavioral and Social Sciences and Education. Washington, DC: National Academy Press.

O'Hare, W. P. (2014). State-level 2010 census coverage rates for young children. *Population Research and Policy Review, 33*(6), 797–816.

O'Hare, W. P. (2015). *The undercount of young children in the U.S. decennial census*. Springer Publishers.

O'Hare, W. P., Robinson, J. G., West, K., & Mule, T. (2016). Comparing the U.S. decennial census coverage estimates for children from the demographic analysis and coverage measurement surveys. *Population Research and Policy Review, 35*(5), 685–704.

Price, D. O. (1947). A check on the underenumeration in the 1940 census. *American Sociological Review, 12,* 44–49.

Robinson, G. J., Bashir, A., & Fernandez, E. W. (1993). Demographic analysis as an expanded program for early coverage evaluation of the 2000 census. *1993 Annual Research Conference*, March 21–24, Arlington VA.

Robinson, J. G. (2000). *Accuracy and coverage evaluation: Demographic analysis results* (p. 1). U.S. Decennial Census Bureau, DSSD Decennial Census 2000 procedures for operations Memorandum Series B-4, U.S. Decennial Census Bureau.

Robinson, J. G. (2010). Coverage of population in decennial census 2000 based on demographic analysis: The history behind the numbers. Decennial Census Bureau, Working Paper No. 91. Available online at http://www.Census.gov/population/www/documentation/twps0091/twps0091.pdf.

Schachter, J. (2008). Estimating native emigration from the United States. Memorandum date December 24, delivered to the U.S. Census Bureau.

Shores, R. (2002). Accuracy and coverage evaluation revision ii adjustment for correlation bias. DSSD, A.C.E. Revision II Memorandum Series PP-53, U.S. Bureau of the Census, Dec.

Shores, R., & Sands, R. (2003). Correlation bias estimation in the accuracy and coverage evaluation revision II. In *Proceedings of the Survey Research methods Section, Joint Statistical Meetings*.

Siegel, J. S., & Zelnik, M. (1966). An evaluation of coverage in the 1960 decennial census of population by techniques of demographic analysis and by composite methods. In *Proceedings of the Social Statistics Section of the American Statistical Association* (1966) (pp. 71–85). Washington, D.C.: American Statistical Association.

United Nations. (1970). *Principles and recommendations for the 1970 population censuses*, Statistical Papers, Series M, No. 44, United Nations, New York, NY.

U.S. Census Bureau. (2003). Technical assessment of A.C.E. revision II, U.S. Decennial Census Bureau, March 12.

U.S. Census Bureau (2004). *Coverage Improvement in the Census 2000 Enumeration*, Census 2000 Topic Report No. 10, Census 2000 Testing, Experimentation, and Evaluation Program , TR-10., March, Washington, DC: U.S. Census Bureau.

U.S. Census Bureau. (2010). The development and sensitivity analysis of the 2010 demographic analysis estimates. In *Population Division Background paper of DA Conference Dec 6, 2010*. 11/29/2010, Table 2. Washington, DC: U.S. Census Bureau.

U.S. Census Bureau. (2012a). *Documentation for the revised 2010 demographic analysis middle series estimates*. Washington, DC: U.S. Census Bureau.

U.S. Census Bureau. (2012b). *2010 census coverage measurement estimation report: Summary of estimates of coverage for persons in the United States*. DSSD 2010 CENSUS COVERAGE MEASUREMNET MEMORANDUM SERIES #2010-G-01. Washington, DC: U.S. Census Bureau.

U.S. Census Bureau. (2012c). *2010 census coverage measurement estimation report: Adjustment for correlation bias*. DSSD 2010 Census Coverage Measurement Memorandum Series #2010-G-11. Washington, DC: U.S. Census Bureau.

U.S. Census Bureau. (2012d). *"2010 census coverage measurement estimation report" net coverage for household population in the United States*. DSSD 2010 Census Coverage Measurement Memorandum Series #2010-G-03. Washington, DC: U.S. Census Bureau.

U.S. Census Bureau. (2014). *Final task force report: Task force on the undercount of young children*. Memorandum for Frank A. Vitrano, February 2. Washington, DC: U.S. Census Bureau.

U.S. Census Bureau. (2016). Investigating the 2010 undercount of young children—A new look at the 2010 census omissions by age. U.S. Census Bureau, 2020 Census Memorandum Series, July 26.

U.S. Census Bureau. (2017). 2020 census decision to change the name of the coverage measurement survey to the post-enumeration survey, Memo for the Record, September 14.

U.S. Office of Management and Budget. (1997). *Revisions to the standards for the classification of federal data on race and ethnicity*. Statistical Policy Directive 15, Federal Register Notice October 30, 1997. Available online at http://www.Whitehouse.gov/omb/fedreg_1997standards.

Velkoff, V. (2011). *Demographic evaluation of the 2010 census*. Paper presented at the 2011 PAA annual conference, Washington, DC.

Wachter, K. W., & Freedman, D. A. (1999). The fifth cell: Correlation bias in U.S. census adjustment. Technical Report Number 570, Department of Statistics, University of California, Berkeley.

West, K. (2012). *Using medicare enrollment file for the DA 2010 estimates*. Paper presented at the Applied Demography Conference at the University of Texas at San Antonio Texas, January.

Wolter, K. (1986). Some coverage error models for census data. *Journal of the American Statistical Association, 81,* 338–353.

Open Access This chapter is licensed under the terms of the Creative Commons Attribution 4.0 International License (http://creativecommons.org/licenses/by/4.0/), which permits use, sharing, adaptation, distribution and reproduction in any medium or format, as long as you give appropriate credit to the original author(s) and the source, provide a link to the Creative Commons license and indicate if changes were made.

The images or other third party material in this chapter are included in the chapter's Creative Commons license, unless indicated otherwise in a credit line to the material. If material is not included in the chapter's Creative Commons license and your intended use is not permitted by statutory regulation or exceeds the permitted use, you will need to obtain permission directly from the copyright holder.

Chapter 4
The Big Picture; Fundamentals of Differential Undercounts

Abstract The 2010 Census coverage error for the total population was very small by international and historic standards, but that masks some large coverage differences among groups. Basic differences in Census coverage by age, sex, race, Hispanic Origin, and tenure are explored in this Chapter. Key findings include, young children (age 0–4) have a higher net undercount than any other age group; males have a higher net undercount than females; Blacks, American Indians on reservations, and Hispanics have relatively high net undercounts; and renters have higher net undercount rates than homeowners. This Chapter is meant to provide an overview and foundation for much of the rest of the book.

4.1 Introduction

When the results of the 2010 Decennial Census were compared to the expected population based on Demographic Analysis (DA) and Dual-Systems Estimates (DSE), the difference was near zero (Velkoff 2011; U.S. Census Bureau 2012). The 2010 Census coverage rate is very good by historical and international standards.

However, the small difference between the expected population and the 2010 Census count for the total population has caused some people to conclude there was a high level of accuracy for the 2010 Census. For example, Rebecca Blank, then Deputy U.S. Commerce Secretary (2012, p. 1) said, "Today's Census Bureau report shows that not only was the 2010 Census delivered on time and significantly under budget- but even more important, it was extremely accurate." A similar sentiment was expressed by Census Bureau Director Groves (2012).

The high level of accuracy for the total population, however, masks crucial differences among demographic groups and that is the point of this book. It is the differential undercounts which are important for many uses of Census data discussed in Chap. 2. If all groups and all places were undercounted at the same rate, the undercount would hardly be a problem. For one thing, we could just inflate the number based on the net undercount rate.

The fundamental problem with differential Census coverage rates is that certain groups have higher net undercount rates than others. This Chapter provides

© The Author(s) 2019
W. P. O'Hare, *Differential Undercounts in the U.S. Census*,
SpringerBriefs in Population Studies, https://doi.org/10.1007/978-3-030-10973-8_4

an overview of key data on Census coverage differentials by age, sex, race, Hispanic Origin, and tenure, to give readers a broad picture of Census undercounts and overcounts. The information in this Chapter will provide readers with data on key measures of Census coverage and establish a foundation for most of the remaining Chapters in this book. All the topics covered in this Chapter will be pursued in more detail in subsequent Chapters.

To keep this Chapter to a manageable size, I look primarily at data from the 2010 Census. Wherever it is feasible, I look at both DA and DSE results. In some cases, however, I draw only on data from DA and in other cases I draw only on data from DSE because one source is much better than the other for a particular population (for example, DA is much better for measuring the net undercount of young children) or data for a particular group are only available from one source (for example, DSE is the only source of information for many race groups). To provide a succinct portrait, only net undercounts and net overcounts are examined in this Chapter, while subsequent Chapters will look at omissions as well as net undercounts.

Data from the 2010 DSE analysis (the method was called Census Coverage Measurement or CCM in the 2010 Census) includes five demographic categories; age, sex, race, Hispanic Origin status, and tenure. Data from the 2010 DA analysis include detailed data by age and sex, but the data on race and Hispanic Origin are limited. Historically the only race groups in DA were Black and Non-Black. In the 2010 Census, the DA data are available for Black Alone and Black Alone or in Combination for the population under age 30. In 2010, DA figures for Hispanics are only available for the population under age 20. For almost all groups the DA results and the DSE results are similar. The major exception to that is for young children (see Table 3.1).

The DSE estimates are based on a sample and therefore include sampling error. So, differences from zero need to be tested to see if the observed differences from zero might be due to chance or random error. Statistical significance is based largely on the size of the difference from zero and the size of the sample on which the estimate is based. Larger differences and estimates from larger samples are more likely to be statistically significant. The statistical significance level used throughout this book is the same one used by the Census Bureau, namely 0.10. This means an observed result that is statistically significant would only happen by random chance alone one time out of ten.

4.2   Census Coverage Differentials by Age

The Census coverage estimates from DA are the best source for examining age differentials for at least four reasons. First, DA data on age are more detailed than that from Dual-Systems Estimates (DSE). Data are available by single year of age from DA but only for large age/sex groups from DSE. Second, DA estimates have been produced since 1950, so there is more historical data. Third, in the decade prior to the 2010 Census, staff at the Census Bureau investigated several issues related to the production of DA estimates (Robinson 2010; Bhaskar et al. 2010; Devine et al.

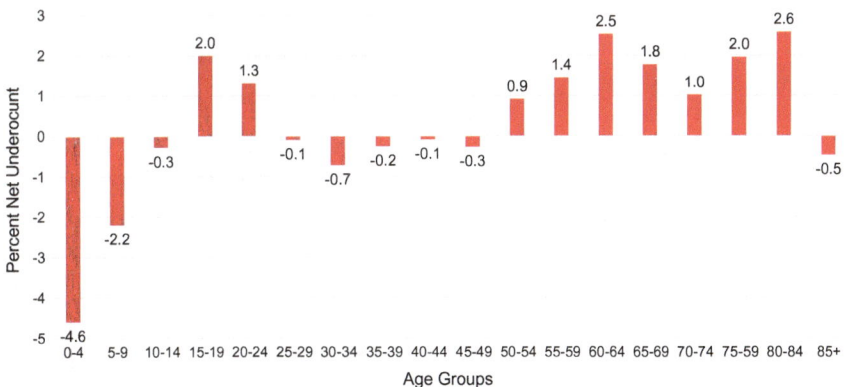

Fig. 4.1 Net undercount rates in 2010 census by five-year age groups. *Source* U.S. Census Bureau, May 2012 DA Release

2010). The increased input review and examination enhance the reliability of the 2010 DA estimates. Fourth, DSE estimates for the youngest ages greatly underestimate the net undercount (O'Hare et al. 2016a, b) so that data series cannot be used to examine the whole age spectrum.

Figure 4.1 shows the net coverage in the 2010 Census by five-year age groups based on DA. Figure 4.1 shows high net undercounts for the youngest population, those age 0–4, and somewhat smaller net undercounts for age 5–9. Young children had a higher net undercount than any other age group in the 2010 Census. The net undercount rate for age 0–4 is 4.6% and for age 5–9 it is 2.2%. No other age group has a net undercount rate of more than 0.7%. The high net undercount of young children has been noted in several recent publications (O'Hare 2014a, b, 2015; U.S. Census Bureau 2014b).

Figure 4.1 also shows relatively high net overcounts for two broad age-groups. The first group is those age 15–24. The second age group that has a net Census overcount is the population age 60 plus. Some of the potential explanations for undercounts and overcounts will be discussed in Chap. 5 along with a more detailed examination of differentials by age.

4.3 Census Coverage Differentials by Sex

There was a net overcount of 1.1% for females and a net undercount of 0.8% for males. As readers will see in Chap. 6, the overall coverage rates for males and females mask large differences across age and race groups. The gender differences in Census coverage are most pronounced in working-age adults.

The 2010 DSE analysis did not provide data on undercounts of males and females for the total population (all ages) but it provided information on males and females

Table 4.1   2010 census net undercount rates from by age and sex

Age	Males	Females
	Rate	Rate
18–29	**−1.2**	0.3
30–49	**−3.6**	**0.4**
50+	**0.3**	**2.4**

Source U.S. Census Bureau (2012, Table 12)
Negative sign reflects a net undercount. The signs here are reversed from the source report in order to keep directionality consistent within this publication
Figures in **BOLD** are statistically significantly different from zero

for three age groups over age 18. The results are shown in Table 4.1. The coverage differentials by sex from DSE are relatively consistent with the results of DA. Data from DSE show males age 18–29 and age 30–49 had net undercount rates that were statistically significantly different than zero while females experienced a net overcount in every age group examined here (although the rates for age 18–29 was not statistically significant). The net undercount estimate for males age 30–49 is particularly high at 3.6%. Both males and females age 50 and over have a net overcount.

4.4   Census Coverage Differentials by Race

I think it is fair to say the Census net undercount differentials for race and Hispanic Origin groups are among the most widely discussed, most well-known, and most contentious. The material in this section provides a brief overview of this issue.

U.S. Decennial Census net undercounts are typically examined through a comparative lens. Historically the comparative paradigm used most often has been the comparison of Black and Non-Black populations (Fein 1989; West et al. 2014; Prewitt 2003; Schwede et al. 2015). In large part, this perspective was based on data that has been available historically. Until recently, Blacks were the only race group identified consistently across states and over time in birth and death certificates data needed for DA estimates. But in many ways, the Blacks/Non-Black comparison also reflects the prevailing Black/White perspective on racial inequality that dominated the 1950s,1960s, and 1970s. This publication expands that perspective to look at differential undercounts for other race groups as well.

For people not familiar with how the federal government collects data on race and Hispanic Origin, this arena can be confusing. Race and Hispanic Origin are measured in the U.S. Census as dictated by U.S. Office of Management and Budget (1997). Categorizing people by race has been a complicated part of the U.S. Census, ever since it was decided to count slaves as 3/5th of a person in the first Census in 1790.

Prewitt (2013) provides a detailed history of how the concepts and measurement of race (and Hispanic Origin) have changed since 1790 in the context of the Census.

Race and Hispanic Origin are two separate concepts in the definitions used by the federal government. In the Census, respondents are first asked if they are Hispanic or not, then they are asked about what race group(s) they identify with.

Starting in 1997, people have been allowed to mark as many race groups as they feel apply. For people who mark only one race, this is referred to as "race alone" for example Black Alone or Asian Alone. People who mark more than one race are added to the "race alone" group for form "Race Alone or in Combination"; for example, "Black Alone or in Combination." In this construction, people can be in more than one group. Someone who marked both Black and White would be included in both Black Alone or in Combination and the White Alone or in Combination but not in the figures for Black Alone or White Alone.

Generally, the 'race alone or in combination" categories are used in this book, since that is the most inclusive way of identifying groups. It is also consistent with the recommendation the U.S. Office of Management and Budget which suggests that when data are used in a civil rights context, the race alone or in combination is the best grouping to use (U.S. Office of Management and Budget 2001). For many minority groups, the Census is largely seen in the context of civil rights enforcement (Leadership Conference Education Fund 2017).

The DSE net undercount rates for different races are shown in Table 4.2. Data for the Non-Hispanic White Alone population is provided because the rates for minority groups are typically contrasted to the rate for the Non-Hispanic White Alone population to develop differential net undercounts. The Non-Hispanic White Alone population is the numerical majority population in the nation.

In the 2010 Census, the Non-Hispanic White Alone population had a net overcount of 0.8%, and most of the other races had net undercounts. The only exception is American Indians Alone or in Combination living in an American Indian Area off reservations where there is a net overcount of 3.9%. In part, because this is such a small population, the net overcount rate for American Indians living off reservations is not statistically significant.

For Blacks Alone or in Combination the net undercount rate is 2.1% in the 2010 Census. There is a 2.9 percentage point difference between the net undercount rate for Blacks Alone or in Combination and the net overcount of Non-Hispanic Whites Alone noted above. This differential is not new. Census coverage data for Blacks are explored in more detail in Chap. 8.

For Asians Alone or in Combination the net undercount is zero. More data for Asians are presented in Chap. 9.

The largest net undercount in Table 4.2 is for American Indians Alone or in Combination on reservations at 4.9%. Overall, American Indian and Alaskan Natives Alone or in Combination had a net undercount of 0.2% which was not statistically significantly different than zero. More information on this population is presented in Chap. 10.

Table 4.2 2010 census estimates of percent net undercount by race (alone or in combination) and Hispanic origin

	Net undercount rate
U.S. total	0.0
Non-Hispanic White alone	**0.8**
Race Alone or in Combination	
Black	**−2.1**
Asian	0.0
American Indian and Alaskan Native	−0.2
On Reservation	**−4.9**
In American Indian Area Off Reservation	3.9
Balance of U.S.	0.1
Native Hawaiian and Pacific Islander	−1.0
Hispanic	**−1.5**

Source U.S. Census Bureau (2012, Table 8)

Except for the Non-Hispanic White Alone category, the race groups include Hispanics

A negative sign reflects an undercount. The signs here are reversed from the source report in order to keep directionality consistent within this publication

Figures in **BOLD** are statistically significantly different from zero

For Native Hawaiians or Pacific Islanders Alone or in Combination the net undercount rate is 1.0% which is not statistically different from zero. More information on this population are provided in Chap. 11.

The Race groups in DA are much more limited than in DSE, but the findings for Blacks and Non-Blacks are very similar to those of DSE. Based on DA analysis, Velkoff (2011) shows a Black Alone or in Combination undercount of 2.5% in the 2010 Census compared to a net overcount of 0.5% for the Non-Black population. The DA method provides data comparing Blacks and Non-Blacks for each Census from 1950 onward and these are presented in Chap. 8.

4.4.1 Hispanics

Table 4.2 shows the net undercount rate for Hispanics was 1.5% in the 2010 Census compared to a net overcount of 0.8% for Non-Hispanic White alone. More data on Census coverage for Hispanics are provided in Chap. 7.

The Hispanic population is treated as homogeneous group with respect to undercount estimates from the Census Bureau. However, it is important to recognize that subgroups within the Hispanic Populations may well have Census coverages rates that are different than the group as a whole (O'Hare 2017). The same is true for Asians. This idea is addressed in more detail in Chaps. 7 and 9.

Table 4.3 2010 census coverage rates by tenure

	Net undercount rate
Population living in owner-occupied housing	**0.6**
Population living in renter-occupied housing	**−1.1**

Source U.S. Census Bureau (2012, Table A)
A negative sign reflects an undercount. The signs here are reversed from the source report in order to keep directionality consistent within this publication
Figures in **BOLD** are statistically significantly different from zero

4.5   Census Coverage Differentials by Tenure

Table 4.3 shows the 2010 Census coverage rates for the population living in owner-occupied housing units and the population living in renter-occupied housing units. There is a net overcount for owners (0.6%) and a net undercount for renters (1.1%). Both figures are statistically significant.

This pattern of Census coverage for homeowners and renters is long standing. To some extent, homeownership may reflect socio-economic status. More information about Census coverage by tenure are presented in Chap. 12.

4.6   Other Groups Missed in the Census

Although Demographic Analysis and Dual-Systems Estimates are the best estimates available for coverage in the U.S. Census, there are some important limitations. For example, the groups for which these two programs produce data are limited. The Census Bureau research only measures Census coverage for five demographic characteristics including;

• Age
• Gender
• Race
• Hispanic Origin Status
• Tenure.

There are other groups which are widely believed to be undercounted in the Census but for which there are no direct undercount estimates from the Census Bureau. Some of these groups are identified by former Under-Secretary of Commerce following the release of data from the 2010 Census. In response to the 2010 Census results being release, then Under-Secretary of Commerce Blank (2012) stated,

> More work remains to address persistent causes of undercounting, such as poverty, mobility, language isolation, low levels of education, and general awareness of the survey.

Table 4.4 Lists of hard-to-count populations from three sources

• Racial and ethnic minorities	• Irregular housing units	• Single and divorced males
• Persons who do not speak English fluently	• Lack of cooperation/trust	• Recent migrants
• Lower income persons	• Communication/Language	• Unemployed
• Homeless persons	• Renters	• Minority ethnic groups
• Undocumented immigrants	• Socio-Economic Status	• Private renters
• Young mobile persons	• Residential mobility	• Those who share a dwelling with other households or with a business
• Children	• Non-City style/non-traditional addresses"	
• Persons who are angry at and/or distrust the government		
• LGBT persons		
Source U.S. Census Bureau's National Advisory Committee on Racial, Ethnic, and Other Population, presentation by Julie Dowling May 27, (2016, page 2)	*Source* Bruce and Robinson (2007)	*Source* Simpson and Middleton (1997)

The factors mentioned by Dr. Blank are commonly believed to be related to Census undercounts, but the Census Bureau does not produce undercount estimates for any of these characteristics (poverty, mobility, language, education, or awareness of the Census).

Groups with high net undercount rates are often referred to as hard-to-count (HTC) populations. According to the U.S. Census Bureau (2016, p. 2);

> Hard-to-count populations face physical, economic, social, and cultural barriers to participation in the Census, and require careful consideration as part of a successful communications strategy.

It is important to recognize that just because the Census Bureau does not produce net undercount estimates for a group that does not mean they are not undercounted. While there are no Census undercount rates for many HTC groups other data such as Mail Return Rates (Leadership Conference Education Fund 2017) or the percent living in hard-to-count areas are sometimes available to shed light on potential Census coverage issues (U.S. Census Bureau 2014a).

Three different lists of hard-to-count populations and/or factors associated with being hard-to-count are presented in Table 4.4. These lists are meant to provide a flavor for the kinds of groups that are often thought to be missed at high rates in the Census.

Table 4.5 Hard-to-count groups compiled by U.S. General Accountability Office based on Perusal of Census Bureau documents

Complex households including those with blended families, multi-generational, or non-relatives
Cultural and linguistic minorities
Displaced persons affected by a disaster
Lesbian gay bisexual transgender queer/questioning persons
Low income persons
Persons experiencing homelessness
Persons less likely to use the Internet and others without Internet access
Persons residing in places difficult for enumerators to access, such as buildings with strict doormen, gated communities, and basement apartments
Persons residing in rural or geographically isolated areas
Persons who do not live in traditional housing
Persons who do not speak English fluently (or have limited English proficiently)
Persons who have distrust of the government
Persons with mental and/or physical disabilities
Persons without a high school diploma
Racial and ethnic minorities
Renters
Undocumented immigrants (or recent immigrants)
Young children
Young, mobile persons

Source U.S. General Accountability Office (2018, Table 1)

Table 4.5 shows a list of hard-to-count populations compiled by the U.S. General Accountability Office based on looking over many Census Bureau documents. Clearly there are many groups that are thought be hard-to-count where the net undercount and omissions are not calculated by the Census Bureau.

4.7   Summary

Several groups have been identified as having relatively high net Census undercounts rates. Groups with the highest net undercount rates include:

- The net undercount for young children (age 0–4) in the 2010 Census was 4.6% which is higher than any other age group.
- Males have a net undercount (0.8%) and females have a net overcount (1.1%).
- Renters have a net undercount (1.1%) while homeowners have a net overcount (0.6%).
- Race and Hispanic Origin Groups with the largest net undercounts are;

– Black Alone or in Combination (2.1%),
– Hispanics (1.5%),
– American Indian Alone or in Combination on reservations (4.9%).

Each of these differential undercounts will be explored in more detail in subsequent Chapters.

References

Bhaskar, R., Scopilliti, M., Hollman, F., & Armstrong, D. (2010). Plans for producing estimates of net international migration for the 2010 demographic analysis estimates. Census Bureau Working Paper No. 90.

Blank, R. (2012, May 22). Statement by Deputy U.S. Commerce Secretary Rebecca Blank on release of the data measuring census accuracy.

Bruce, A., & Robinson, J. G. (2007). *Tract-level planning database with census 2000 census data*. Washington, DC: U.S. Census Bureau.

Devine, J., Sink, L., DeSalvo, B., & Cortes, R. (2010). The use of vital statistics in the 2010 demographic analysis estimates. Census Bureau Working Paper No. 88,

Fein, D. J. (1989). *The social sources of census omission: Racial and ethnic differences in omission rates in recent U.S. censuses*. Dissertation in Department of Sociology, Princeton University, Princeton, NJ.

Groves, R. (2012, May 30). How good was the 2010 census? A view from the post-enumeration survey, Directors Blog.

Leadership Conference Education Fund. (2017). Fact sheet: The census and civil rights. Downloaded on June 13, 2017 from http://civilrightsdocs.info/pdf/Census/Fact-Sheet-Census-and-Civil-Rights.pdf.

O'Hare, W. P. (2014a). Assessing net coverage error for young children in the 2010 U.S. decennial census. *Center for Survey Measurement Study Series (Survey Methodology #2014-02)*. U.S. Census Bureau. Available online at http://www.Census.gov/srd/papers/pdf/ssm2014-02.pdf.

O'Hare, W. P. (2014b). Historical examination of net coverage error for children in the U.S. decennial census: 1950 to 2010. *Center for Survey Measurement Study Series (Survey Methodology #2014-03)*. U.S. Census Bureau. Available online at http://www.Census.gov/srd/papers/pdf/ssm2014-03.pdf.

O'Hare, W. P. (2015) *The undercount of young children in the U.S. Decennial Census*. Springer Publishers.

O'Hare, W. P., Robinson, J. G., West, K., & Mule, T. (2016a). Comparing the U.S. decennial census coverage estimates for children from the demographic analysis and coverage measurement surveys. *Population Research and Policy Review, 35*(5), 685–704.

O'Hare, W. P., Mayol-Garcia, Y., Wildsmith, E., & Torres, A. (2016b). The invisible ones: How Latino children are left out of our nation's census count. Child Trends Hispanic Institute & National Association of Latino Elected Officials, Child Trends, Washington DC.

O'Hare, W. P. (2017). *Counting all Californians in the 2020 census*. The Census Project. https://Censusproject.files.wordpress.com/2017/10/calif-report-10-16-2017_format-final.pdf.

Prewitt, K. (2003). *Politics and science in census taking*, in series The American People: Census 2000, Russell Sage Foundation and Population Reference Bureau, Population Reference Bureau, Washington, DC.

Prewitt, K. (2013). *What is your race: The census and our flawed efforts to classify Americans*. Princeton, NJ: Princeton University Press.

Robinson, J. G. (2010). Coverage of population in census 2000 based on demographic analysis: The history behind the numbers. U.S. Census Bureau, Working Paper No. 91.

Schwede, L., Terry, R., & Hunter, J. (2015). Ethnographic evaluations on coverage of hard-to-count minority in the US decennial censuses. In R. Tourangeau, B. Edwards, T. P. Johnson, K. M. Wolter, & N. Bates (Eds.), *Hard-to-survey populations* (pp. 293–315). Cambridge, England: Cambridge University Press.

U.S. Census Bureau. (2012). *2010 census coverage measurement estimation report: Summary of estimates of coverage for persons in the United States*. DSSD 2010 Census Coverage Measuremnet Memorandum Series #2010-G-01. Washington, DC: U.S. Census Bureau.

U.S. Census Bureau. (2014a). Planning Data Base. Available at https://www.census.gov/research/ data/planning_database/2014/.

U.S. Census Bureau. (2014b, February 2). *Final task force report: Task force on the undercount of young children*. Memorandum for Frank A. Vitrano. Washington, DC: U.S. Census Bureau.

U.S. Census Bureau. (2016). *Developing an integrated communication strategy: Select topics in international census*. Washington, DC: U.S. Census Bureau.

U.S. General Accountability Office. (2018, July). 2020 census: Actions needed to address challenges to enumerating hard-to-count groups, GAO-18-599.

U.S. Office of Management and Budget. (1997, October 30). *Revisions to the standards for the classification of federal data on race and ethnicity*. Statistical Policy Directive 15, Federal Register Notice.

U.S. Office of Management and Budget. (2001). *Guidance on aggregation and allocation of data on race for use in civil rights monitoring and enforcement*.

Velkoff, V. (2011). *Demographic evaluation of the 2010 census*. Presentation at the Population Association of America Annual Conference.

West, K., Devine, J., & Robinson, J. G. (2014). *An assessment of historical demographic analysis estimates for the Black male cohorts of 1935–39*. Paper presented at the Annual Meeting of the American Statistical Association, Boston, MA.

Simpson, L., & Middleton, E. (1997). *Who is missed by a National Census? A review of empirical results from Australia, Britain, Canada, and the USA* (p. 1). The Cathie Marsh Centre for Census and Survey Research, University of Manchester, UK.

Open Access This chapter is licensed under the terms of the Creative Commons Attribution 4.0 International License (http://creativecommons.org/licenses/by/4.0/), which permits use, sharing, adaptation, distribution and reproduction in any medium or format, as long as you give appropriate credit to the original author(s) and the source, provide a link to the Creative Commons license and indicate if changes were made.

The images or other third party material in this chapter are included in the chapter's Creative Commons license, unless indicated otherwise in a credit line to the material. If material is not included in the chapter's Creative Commons license and your intended use is not permitted by statutory regulation or exceeds the permitted use, you will need to obtain permission directly from the copyright holder.

Chapter 5
Census Coverage Differentials by Age

Abstract People pass through different family situations and living arrangements as they age and many of these changes over a lifetime are related to changes in the likelihood of being missed in the Census. In this Chapter, some of the biggest differences in Census coverage by age are examined and some ideas about why people are missed or overcounted are explored. Young children age 0–4 had the highest net undercount rate and highest omissions rate of any age group in the 2010 Census. The college-age population (age 18–24) and elderly people (over age 60) had net overcounts.

5.1 Introduction

In the 2010 Census, some age groups had net undercounts, some had net overcounts, and many age groups experienced little net coverage error. In this Chapter, the focus is on the age groups that had the largest net undercounts and net overcounts. In addition to looking at the net undercounts and overcounts of age groups, omissions rates are also examined.

The initial analysis in this Chapter relies heavily on Demographic Analysis (DA) estimates. I believe the strengths of DA methodology make it a particularly good technique for discussing Census coverage by age for at least four reasons. First, DA data on age is more detailed than that from Dual-Systems Estimates (DSE). Data are available by single year of age from DA but only for large age/sex groups from DSE. Second, DA estimates have been produced since 1950, so there is more historical data. Third, in the decade prior to the 2010 Census, staff at the Census Bureau investigated several issues related to the production of DA estimates (Robinson 2010; Bhaskar et al. 2010; Devine et al. 2010). The increased input, review, and examination enhance the reliability of the 2010 DA estimates. Fourth, DSE estimates for the youngest ages greatly underestimate the net undercount (O'Hare et al. 2016) so that data series cannot be used to examine the whole age spectrum.

© The Author(s) 2019
W. P. O'Hare, *Differential Undercounts in the U.S. Census*,
SpringerBriefs in Population Studies, https://doi.org/10.1007/978-3-030-10973-8_5

5.2   Census Net Undercounts by Age

Figure 5.1 shows net coverage in the 2010 Census by five-year age groups based on DA. Data in Fig. 5.1 indicate a high net undercount for young children, a high net overcount for people in their late teens and early twenties and a high net overcount for people over age 60. For other age groups the net coverage is close to zero (less than 1%). Therefore, I focus on three groups, age 0–4, 18–24 and 60 and older in the remainder of this Chapter.

Perhaps the most surprising finding shown in Fig. 5.1 is the high net undercount of young children. In the words of former Census Bureau Director Groves (2010, p. 1).

> It's often a surprise to many people when they learn that children tend to be undercounted in the US Censuses. Most can imagine various types of adults who fail to participate in Censuses, but don't immediately think of children being missed.

Moreover, the survey research literature shows that households with children generally respond to surveys at higher rates than those without children (Groves and Couper 1998; Brick and Williams 2012). Groves and Couper (1998, p. 138) offer this succinct summary of the relationship between children in the household and cooperation in survey research, "Without exception, every study that has examined response or cooperation finds positive effects of the presence of children in the household."

Nonetheless, Fig. 5.1 shows the population age 0–4 had the highest net undercount (4.6%) of any age group in the 2010 Census. There was a somewhat smaller net undercount (2.2%) for age 5–9. No other age group had a net undercount rate of more than 0.7%. The net undercount of young children is not only the largest net undercount of any age group, it is the largest Census coverage error in either direction (i.e. net undercounts or net overcounts). The high net undercount of young children

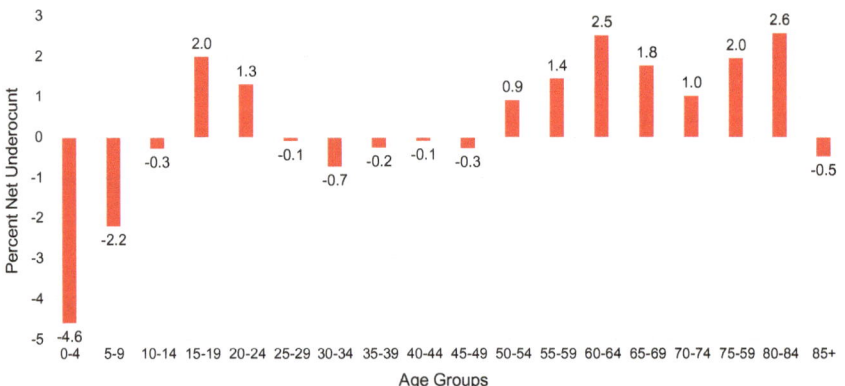

Fig. 5.1 Net undercount rates in 2010 census by five-year age groups. *Source* U.S. Census Bureau, May 2012 DA Release

has been noted in several recent publications (O'Hare 2014a, b, 2015; U.S. Census Bureau 2014).

Consistent with much of the other literature on Census undercounts, among young children, racial and ethnic minorities had higher net undercounts. Based on the 2010 Census DA release, there was a net undercount of 6.3% for Black Alone or in Combination age 0–4 and 7.5% for Hispanics age 0–4 (O'Hare 2015). It should be noted that a couple of recent studies (King et al. 2018; Jensen et al. 2018) suggest that the net undercount of young children might not be quite as high as earlier estimates indicate but nonetheless is still higher than any other age group, by far. These new studies are more in the nature of experimental estimates rather than officials estimates.

The high net undercount of young children is not a new issue. Difficulty in enumerating young children accurately has been noted historically (Hacker 2013; Adams and Kasakoff 1991; U.S. Census Bureau 1944). More than 100 years ago, Young (1901, p. 21) stated, "Experience has shown that it is extremely difficult to ascertain the true number of children in any population by simple enumeration." The passage below is from a Census Bureau report following the 1940 Census, (U.S. Census Bureau 1944, p. 32) "Underenumeration of children under 5 years old, particularly infants under one year old, has been uniformly observed in the United States Census and in the Censuses of England and Wales and of various countries of continental Europe." The results of the 2010 U.S. Census suggest this situation has not changed much since then.

Why do young children have such a high net undercount rate? It is widely believed that there is not just one cause for the high net undercount of young children in the Census but there are many causes. With respect to the high net undercount of young children, the Census Bureau Task Force on the Undercount of Young Children (U.S. Census Bureau 2014) concluded, "The task force is convinced that there is no single cause for this undercount." O'Hare (2015, Chap. 7) discusses several potential ideas about why young children have a high net undercount in the Census. Over the past few years, the Census Bureau has engaged in several studies to learn more about the undercount of young children and they have produced a summary of the results (O'Hare et al. forthcoming).

Any explanation of why young children have such a high net undercount must not only explain why young children are missed, but why they are missed as a much higher rate than older children or adults. In some prior analysis, all children age 0–17 have been grouped together. For example, in the 1990 DSE results (U.S. Census Bureau 2001) all children under age 18 were treated as one group. But children (age 0–17) are not homogenous with respect to the risk of being missed in the Census.

Results from the 2010 Census shown in Fig. 5.2 indicate the net undercount is much higher for those age 0–4 than those age 14–17 (the population age 14–17 actually had a net overcount). When young children are grouped with older children it is difficult to discern why younger children are missed and older children overcounted.

Figure 5.2 shows there is almost a perfect correlation between age and Census coverage. What is responsible for this relationship between age and Census coverage? I am not aware of any theory or evidence that has been put forward to explain this strong statistical relationship.

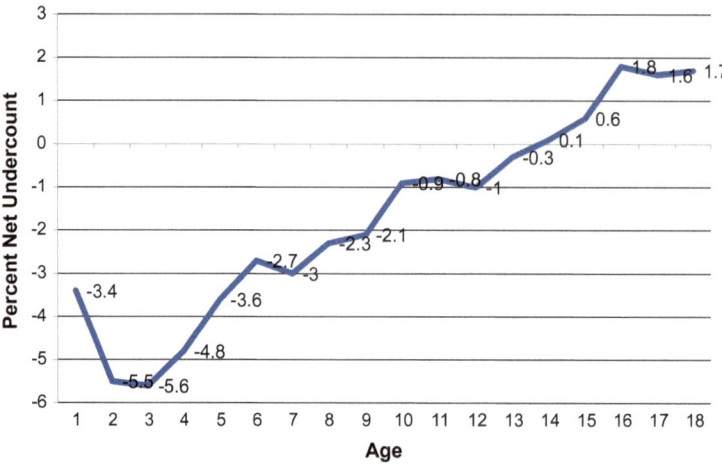

Fig. 5.2 2010 census net undercount rates by single year of age: 0–17. *Source* U.S. Census Bureau, May 2012 DA Release

One factor related to the high net undercount of young children that is increasingly clear is the fact that young children are more highly concentrated than older children in the kinds of households and families that are more difficult to enumerate. Table 5.1 shows the percent of young children (age 0–4) compared to older children (age 10–17) in each of 14 situations thought to be related to being missed in the Census. In nearly every situation, young children are more highly concentrated than older children.

Recent research by Fernandez et al. (2018) shows that among children under age 5, the odds of being missed in the Census are higher for:

- Grandchildren
- Children living in single-parent families
- Children in large (7 or more people) households
- Children in poverty
- Children in households where no adult completed college
- Children in households where one or more persons are unemployed
- Children in immigrant households (half or the people are foreign born)
- Children in households where no one speaks English "well" or better.

Fernandez and her colleagues found that, by and large, these factors hold for Non-Hispanic Whites, Hispanics, and Non-Hispanic Blacks.

Research by Fernandez et al. (2018) also shows that one of the factors that is most closely associated with young children being missed in the Census is whether they were counted in the self-response or Non-Response Followup phase of the Census. Based on logistic regression analysis, Fernandez et al. (2018) show that if young children are not included in the self-response phase of the Census, they are 74% more likely to be missed. This research suggests that the Census Bureau should enhance training among 2020 Census enumerators with respect to making sure

Table 5.1 Hard-to-count characteristics of children by age

	Percent of age group with this characteristics		
	Age 0–4	Age 10–17	Difference (0–4 minus 10–17)
Age of household 18-29[a]	29	3	26
Renter[a]	44	32	12
Not in single detached unit[b]	38	26	12
Household receives cash public assistance or SNAP[b]	31	23	9
Different address one year ago[b]	25	12	13
Complex household[a]	40	33	6
Below poverty[b]	25	19	6
Enumerator completed response[a]	31	27	5
Living in a single parent in poverty[b]	17	13	4
Grandchild of householder[b]	11	5	6
Not Biological or adopted child[b]	16	15	2
Large (6 plus person) household[a]	23	22	1
Households with limited English[b]	26	27	−1
Not born in US[b]	2	7	−5

[a]*Source* 2010 Census, U.S. Census Bureau (2017a)
[b]*Source* U. S. Census Bureau (2017b)

all young children in a household are included on returned Census questionnaires. Recently the Census Bureau (2018) issued a short publication showing some of the main reasons young children are missed in the Census.

5.3 High Net Overcounts of College-Age Population

One group with a relatively high net overcount is the population age 15–24. The net overcount for age 15–19 was 2.0% and for those age 20–24 it was 1.3%. The overcount in this age group is typically attributed to young adults being counted in the home of their parents as well as another address such as a college dormitory or military barracks (Martin 2007). Figure 5.3 shows net undercounts for the population age 15–24 by single year of age and reveals that the net overcounts peaks around age 19 or 20, which is the age many young people are in college.

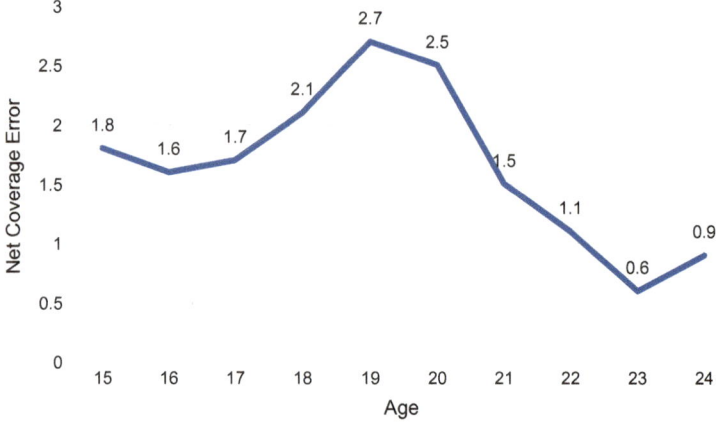

Fig. 5.3 2010 census net coverage error by single year of age: 15–24. *Source* U.S. Census Bureau, May 2012 DA Release

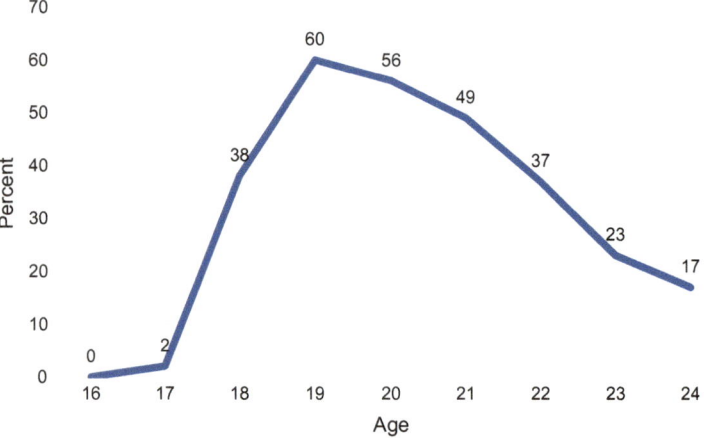

Fig. 5.4 Percent of population in colleges by age: 16–24. *Source* U.S. Census Bureau, American Community Survey

Figure 5.4 shows the percent of people attending college by single year of age from 16–24 based on data from the Census Bureau's American Community Survey. Figure 5.4 shows college attendance peaking at age 19, 20, and 21, the exact same ages as the higher net overcounts. Note that the ACS question just asked about college attendance, it does not differentiate those who leave home to go to college from those who live at home and attend college.

The 2010 Census Coverage Followup operation, showed relatively large duplication errors for people in college and people in jail (U.S. Census Bureau 2012, Table 24.) People age 18–24 are over-represented in both groups.

5.4 Net Overcounts of Elderly Population

The second age group that had a substantial Census net overcount is the population age 60 and over. The overcount for this group may be attributed to retirees with two (or more) homes and to a lesser extent older people in nursing homes or other long-term care facilities. Like young adults, some of these people are counted in more than one place. Williams (2012, p. 8) provides one example of a Census overcount, "A husband and wife, for example, might own a vacation home and fill out a questionnaire there as well as their usual residence." The 2010 Census Coverage Followup operation, showed relatively large duplication errors for people in nursing homes (U.S. Census Bureau 2012, Table 24).

In an analysis following the 2000 Census, the Census Bureau focused on people who had been included in the Census more than once (U.S. Census Bureau 2002). In a key table, they show that people over age 50 are much more likely than younger groups to be duplicated in a different state, which suggests they were being counted in more than one home such as a winter home in Florida and a summer home in Michigan. This idea is bolstered by trends from 1950 to 2010 shown in Sect. 5.6.

5.5 Omissions in the 2010 Census

Recall that the net Census undercount rate is a balance between people omitted and those included erroneously (mostly double counted) and whole-person imputations. The omissions rate captures the share of a group missed in the Census. DSE is the only method that shows omissions rates.

In many ways the omissions rate is a more meaningful statistic because in the net undercount calculation, omissions can be cancelled out by erroneous inclusions or double counting. A net undercount of 0 could be the result of no one missed and no one double counted, or for example, 10% missed, and 10% double counted.

The omissions rates for the population age 0–9 published by the Census Bureau in 2012 are suspect because the DSE methodology only reflects a small portion of the net undercount of young children in the 2010 Census (O'Hare et al. 2016). DA estimated a net undercount of 4.6% for the population age 0–4 compared to only 0.7% for DSE. The difference is generally ascribed to correlation bias in the DSE methodology (O'Hare et al. 2016). Correlation bias refers to the fact that the kinds of people missed in the Census are also missed in the Post-Enumeration Survey.

However, recent research by the Census Bureau's Task Force on the Undercount of Young Children provides updated estimates for omissions rates for several large age/sex groups by taking advantage of the strength of the DA and DSE methods (U.S. Census Bureau 2016). These improved estimates for omissions in the 2010 Census are provided in Table 5.2.

The updated omissions calculations are only available for a few age/sex groups and only for the population as a whole and not any race/Hispanic groups. The rates

Table 5.2 2010 Omissions by Age (Updated in 2016)

	Number of omissions (in 1000s)	Omission rate
Age 0–4	2172	10.3
Age 5–9	1517	7.3
Age 10–17	625	4.9
Males age 18–29	1883	7.9
Females age 18–29	1514	6.4
Males age 30–49	3012	7.3
Females age 30–49	1171	2.8
Males age 50+	1793	4.0
Females age 50	949	1.9
Total	15,636	5.2

U.S. Census Bureau (2016), Table 3

are similar to the omissions rates released by the Census Bureau in 2012 except for two groups. For young children, the updated omissions rates are much higher than the earlier omissions rate estimates from the Census Bureau. For young adults the omissions rate from the updated analysis is somewhat higher than the original DSE analysis.

Young children had the highest net undercount rate of any age group and Table 5.2 shows they had the highest omissions rate of any age group in the 2010 Census. The population age 0–4 had an omissions rate of 10.3% which translate into nearly 2.2 million young children omitted from the 2010 Census.

5.6   Trends Over Time

Examination of net Census coverage rates from 1950 to 2010 indicates a significant and steady reduction in the net undercount in the total population. However, when the overall trend is decomposed by age, a more complex story emerged.

Figure 5.5 shows net undercounts by five-year age groups in every Census since 1950. While this figure is somewhat complicated here are a couple of trends that are clear. The net undercount of young children is long-standing as it has been seen in every Census since 1950. The net undercount of young children varied between 1.4 and 6% between 1950 and 2010, although the net undercount of young children has increased substantially since 1980. In fact, the net undercount of young children in the 2010 Census is almost exactly the same as the net undercount of young children in the 1950 Census.

The net overcount of people age 60 and older has emerged over the past several decades. Figure 5.5 shows the net overcount for this age group was higher in 2010 than any other Census shown and the rates for the 2000 Census are not far behind

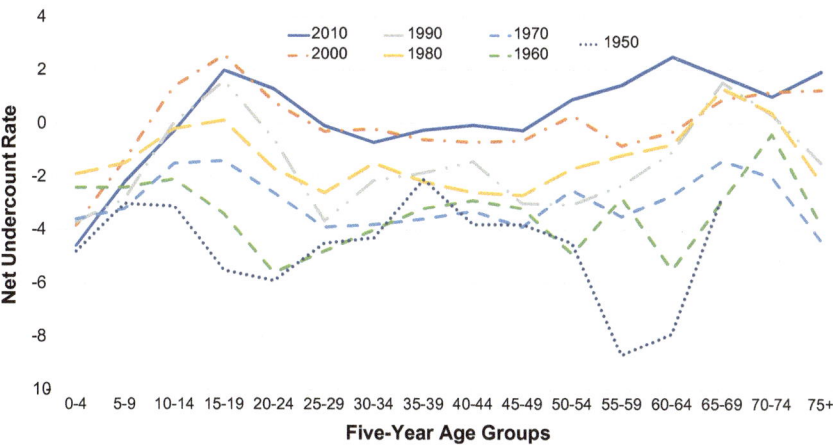

Fig. 5.5 Net census undercount rates by five-year age groups: 1950–2010

those of 2010. On the other hand, the net undercount rates for those age 60 and older in 1950, 1960, and 1970, are relatively low compared to the past two Censuses. This supports the idea that the overcount of the older population may be related to increases in dual home ownership for this age group which increased over this time period.

Figure 5.5 also shows that the coverage of college-aged people has changed over time. In 1950, 1960 and 1970, there was a net undercount for age 15–19 and 20–24, but in the 2000 and 2010 Census, there was a net overcount in these age groups. This is consistent with larger number of people in the young adult age group leaving the home of their parents in recent decades.

One important divergence in net census coverage trends since 1950 may not be clear from Fig. 5.5. Figure 5.6 shows net undercount rates for the adult population (age 18+) and the young child population (age 0–4) for each U.S. Decennial Census from 1950 to 2010. Figure 5.6 shows there are two very distinct periods between 1950 and 2010 related to the net undercount of young children. Between 1950 and 1980, the net undercount rates of both young children and adults improved and the differences between young children and adults were not large. Specifically, the net undercount rates for the adult population went from 3.8% undercount in 1950 to a 1.4% undercount in 1980. While the net undercount for young children fell from 4.7 to 1.4% in the same period.

Following the 1980 Census. The net undercount of young children and adults began diverging. The coverage rates for adults continued the improvement seen in the 1950–1980 period while the net undercount rates for young children increased following 1980. Specifically, the coverage rates for adults went form 1.4% net undercount in 1980 to a 0.7% overcount in 2010. The net undercount for young children increased from 1.4% in 1980 to 4.6% in 2010.

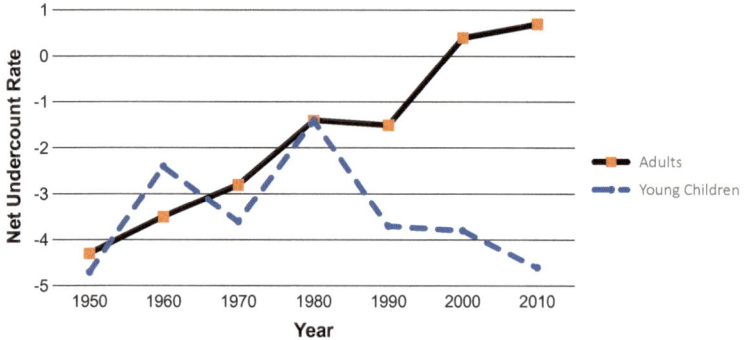

Fig. 5.6 Net undercount rates for young children (0–4) and adults (18+): 1950–2010. *Source* O'Hare (2014a, b)

5.7 Summary

Net undercounts and overcount as well as omissions rates vary by age. Young children age 0–4 had the highest net undercount rate and highest omissions rate of any age group in the 2010 Census. Young Hispanic children and young Black Alone or in Combination children had net undercount rates that are about 50% higher than the overall net undercount rate for this age group. One reason for the high net undercount rate of young children is the fact that they are concentrated in hard-to-count families and households.

The college age population (age 18–24) had a net overcount in the 2010 Census. The high net overcount is attributed to many young adults being counted at their parent's home as well as another location such as college dormitory. This age group also had a relatively high omissions rate.

For the population age 60+, there was a relatively high net overcount in 2010. This is attributed to the fact that many people in this age group have two (or more) homes and are counted in both homes.

References

Adams, J. W., & Kasakoff, A. B. (1991). Estimates of U.S. decennial census underenumeration based on genealogies. *Social Science History*, *15*(4, Winter), 527–543.

Bhaskar, R., Scopilliti, M., Hollman, F., & Armstrong, D. (2010). *Plans for producing estimates of net international migration for the 2010 demographic analysis estimates*. Census Bureau Working Paper No. 90.

Brick, J. M., & Williams, D. (2012). Explaining rising nonresponse rates in cross-sectional surveys. *The Annals of the American Academy of Political and Social Science*.

Devine, J., Sink, L., DeSalvo, B., & Cortes, R. (2010). *The use of vital statistics in the 2010 demographic analysis estimates*. Census Bureau Working Paper No. 88.

Fernandez, L., Shattuck, R., & Noon, J. (2018). *Using administrative records and the American community survey to study the characteristics of undercount young children in the 2010 census.* Working Paper CARRA.

Groves, R. (2010). *Children count too!* Census Bureau's Directors Blog March 9. Washington, DC: U.S. Census Bureau.

Groves, R. M., & Couper, M. P. (1998). *Nonresponse in household interview sur*veys. Wiley.

Hacker, J. D. (2013). New estimates of census coverage in the United States: 1850–1930. *Social Science History*, 37(1, Spring).

Jensen, E., Benetsky, M., & Knapp, A. (2018). *A sensitivity analysis of the net undercounts for young Hispanic children in the 2010 census.* Poster at eh 2018 Population Association of American conference, Denver, CO, April 25–28 downloaded May 5, 2108 at https://paa.confex.com/paa/2018/meetingapp.cgi/Paper/20826.

King, H., Ihrke, D., & Jensen, E. (2018). *Subnational estimates of net coverage error for the population aged 0 to 4 in the 2010 census.* Paper present the 2018 Population Association of American Conference, April 25–28, Denver CO, Downloaded May 6, 2018 https://paa.confex.com/paa/2018/meetingapp.cgi/Paper/21374.

Martin, E. (2007). Strength of attachment: Survey coverage of people with tenuous ties to residences. *Demography, 44*(2), 437–440.

O'Hare, W. P. (2014a). *Assessing net coverage error for young children in the 2010 U.S. Decennial Census.* Center for Survey Measurement Study Series (Survey Methodology #2014-02). U.S. Census Bureau. Available online at http://www.Census.gov/srd/papers/pdf/ssm2014-02.pdf.

O'Hare, W. P. (2014b). *Historical examination of net coverage error for children in the U.S. Decennial Census: 1950 to 2010.* Center for Survey Measurement Study Series (Survey Methodology #2014-03). U.S. Census Bureau. Available online at http://www.Census.gov/srd/papers/pdf/ssm2014-03.pdf.

O'Hare, W. P. (2015). *The undercount of young children in the U.S. Decennial Census.* Springer Publishers.

O'Hare, W. P., Griffin, D., & Konicki, S. (forthcoming). *Investigating the 2020 undercount of young children—Summary of recent research.* U.S. Census Bureau, 2020 Census Memorandum Series.

O'Hare, W. P., Robinson, J. G, West, K., & Mule, T. (2016). Comparing demographic analysis and dual-systems estimates results for children. *Population Research and Policy Review*, 35(5), 685–704.

Robinson, J. G. (2010). *Coverage of population in census 2000 based on demographic analysis: the history behind the numbers.* U.S. Census Bureau, Working Paper No. 91.

U.S. Census Bureau. (1944). *Population differential fertility: 1940 and 1910 standardized fertility rates and reproduction rates, Appendix A, completeness of enumeration of children under 5 years old, U.S. census of 1940 and 1910, 1940 census.* Washington, DC: U.S. Census Bureau.

U.S. Census Bureau. (2001). *ESCAP II: Demographic analysis results.* Executive Steering Committee for A.C.E. Policy II, Report No. 1, J Gregory Robinson, June 13.

U.S. Census Bureau. (2002). DSSD A.C.E. Revision II Memorandum Series #PP-51, A.C.E. Revision II results: Further study of person duplicates. Washington, DC: U.S. Census Bureau.

U.S. Census Bureau. (2012). *2010 census coverage measurement estimation report: Summary of estimates of coverage for persons in the United States.* DSSD 2010 Census Coverage Measurement Memorandum Series #2010-G-01. U.S. Census Bureau: Washington, DC.

U.S. Census Bureau. (2014). *Final task force report: Task force on the undercount of young children.* Memorandum for Frank A. Vitrano, U.S. Census Bureau, Washington, DC. February 2.

U.S. Census Bureau. (2016). *Investigating the 2010 undercount of young children—A new look at 2010 census omission by age*, 2020 Census Memorandum Series. July 2016.

U.S. Census Bureau. (2017a). *Investigating the 2010 undercount of young children—A comparison of demographic, housing and household characteristics of children by age.* January 18.

U. S. Census Bureau. (2017b). *Investigating the 2010 undercount of young children—A comparison of demographic, social, and economic characteristics of children by age.* Scott Konicki, and Deborah Griffin July 7.

U.S. Census Bureau. (2018). *Counting young children in the 2020 census*. Washington, DC: U.S. Census Bureau. Available at https://www.census.gov/content/dam/Census/newsroom/press-kits/2018/counting-young-children-in-2020-census.pdf.

Williams, J. D. (2012). *The 2010 decennial census: Background and issues*. CRS Report for Congress, Congressional Research Services, 7-5700, R4055.

Young, A. A. (1901). The enumeration of children. *Publications of the American Statistical Association, 7*(53), 21–48.

Open Access This chapter is licensed under the terms of the Creative Commons Attribution 4.0 International License (http://creativecommons.org/licenses/by/4.0/), which permits use, sharing, adaptation, distribution and reproduction in any medium or format, as long as you give appropriate credit to the original author(s) and the source, provide a link to the Creative Commons license and indicate if changes were made.

The images or other third party material in this chapter are included in the chapter's Creative Commons license, unless indicated otherwise in a credit line to the material. If material is not included in the chapter's Creative Commons license and your intended use is not permitted by statutory regulation or exceeds the permitted use, you will need to obtain permission directly from the copyright holder.

Chapter 6
Census Coverage Differentials by Sex

Abstract Males and females often play different roles in society at different ages and these can impact family situations and living arrangements related to the likelihood of being missed in the Census. In this Chapter, examination of differences in Census coverage by sex show that males generally have Census net undercount rates while females have net overcounts. Since 1940, the net undercount rates of both males and females have improved a lot but the differential between males and females has not narrowed.

6.1 Introduction

This Chapter focuses on Census coverage for males and females including coverage rates for subgroups defined by age, race, Hispanic Origin, and tenure. In addition to looking at the net undercounts and overcounts of males and females, omissions rates by sex are also examined.

The initial analysis shown in this Chapter relies heavily on Demographic Analysis (DA) estimates. I believe the strengths of DA methodology make it a particularly good technique for discussing Census coverage by age for at least four reasons. First, DA data on age are more detailed than that from Dual-Systems Estimates (DSE). Data are available by single year of age from DA but only for large age/sex groups from DSE. Second, DA estimates have been produced since 1940, so there is more historical data. Third, in the decade prior to the 2010 Census, staff at the Census Bureau investigated several issues related to the production of DA estimates (Robinson 2010; Bhaskar et al. 2010; Devine et al. 2010). The increased input, review, and examination enhance the reliability of the 2010 DA estimates. Fourth, DSE estimates greatly underestimate the net undercount for young children (O'Hare et al. 2016) so that data series cannot be used to examine differential undercounts for the whole age spectrum.

Both methods for measuring Census accuracy (DA and DSE) show males had lower coverage rates than females. In the 2010 Census, the DA method shows a net undercount rate of 0.8% for males compared to a 1.1% net overcount for females. In the 2010 Census, the DSE method also shows adult males typically had net under-

© The Author(s) 2019
W. P. O'Hare, *Differential Undercounts in the U.S. Census*,
SpringerBriefs in Population Studies, https://doi.org/10.1007/978-3-030-10973-8_6

counts while adult females typically had net overcounts. This result is not surprising. Simpson and Middleton (1997) identify adult males as being among the groups most difficult to enumerate in Censuses.

6.2   Undercounts by Sex and Age

As readers will see, in terms of Census coverage, sex matters at some ages but not very much at other ages. Figure 6.1 shows net undercount rates in the 2010 Census for males and females by five-year age groups. In general, there are no differences by sex for the youngest or the very oldest age groups, but there are very noticeable differences among adults from their mid-20s to mid-70s. Females in this age range have better coverage than males. The biggest gaps between males and females are in the 25–45-year old age range, and the gaps seen between age 45 and 65 are somewhat smaller. The biggest single year difference is age 31 where there is a 4.8% net undercount for males and a 0.2% net overcount for females.

Data from DSE are shown in Table 6.1. For age 18–49, there were statistically significant net undercount rates for males but small net overcount rates for females. For the population age 50 and over, there were net overcounts for both groups but the net overcount for males (0.3%) is much smaller than that for females (2.4%).

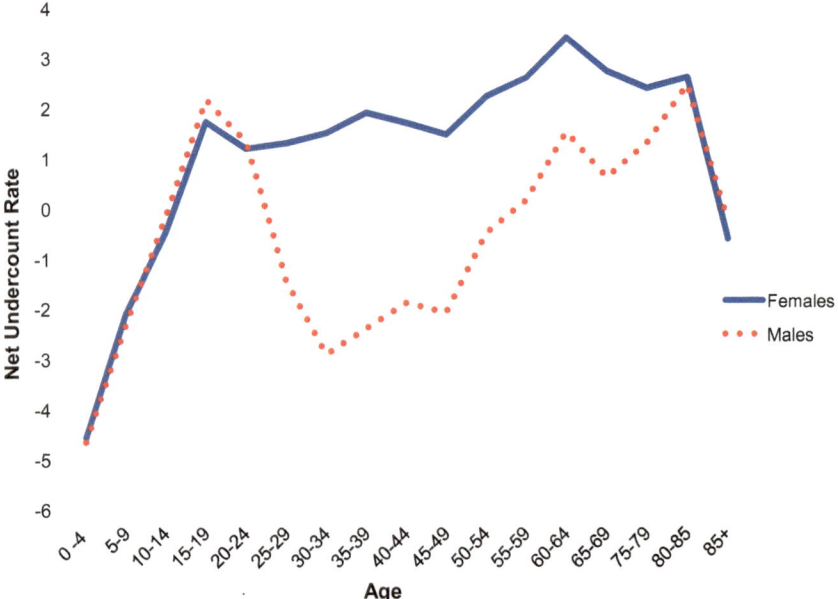

Fig. 6.1 2010 census net undecount rates for males and females by five-year age groups. *Source* U.S. Census Bureau, May 2012 DA Relesae

Table 6.1 2010 census net undercount rates from census coverage measurement by age and sex

Age	Males	Females
18–29	**−1.2**	0.3
30–49	**−3.6**	**0.4**
50 +	**0.3**	**2.4**

Source U.S. Census Bureau (2012b), Table 12
A negative sign reflects a net undercount. The signs here are reversed from the source report in order to keep directionality consistent within this publication
Figures in **BOLD** are statistically significantly different from zero

Table 6.2 Census coverage for males and females by race in the 2010 census

	Males	Females
Total	−0.8	1.1
Black Alone	−4.6	−0.1
Non-Black Alone	−0.3	1.1

Source U.S. Census Bureau, Demographic Analysis tables released May 2012

To some extent these figures may reflect the position of males and females in broader society. For example, it is more acceptable for males and females in their 20s to remain living in their parents' home, so they are apt to have a stable housing situation at that stage in their lives. As they move into their 30s and 40s, it becomes less acceptable to remain living in their parents' home (particularly for males) so their housing situations are more likely to be unstable. Also, many females in their 20s, 30s and 40s, are likely to be responsible for taking care of children on a daily basis which makes stable housing more important. A stable housing situation is associated with a higher likelihood of being counted in the Census (Martin 2007).

6.3 Undercount by Sex and Race

The overall Census coverage figures for males and females mask big differences by race. These differences are first examined using DA data then using data from DSE. Recall that in DA, the only race group that can be identified for all ages is Black and the residual category is called Non-Black.

Table 6.2 shows that based on data from DA, there was a net undercount of 4.6% for Black Alone males compared to a small net undercount (0.1%) for Black Alone females resulting in a 4.5 percentage point gap. For the Non-Black Alone males there was a net undercount of 0.3% compared to a net overcount of 1.1 for females in this population, so the gap is 1.4 percentage points. In other words, males have lower coverage rates than females in both groups, but the differential is much larger for Blacks Alone than for Non-Blacks Alone.

Table 6.3 2010 Census net undercount rates by sex, race, and age

		Age 18–29	Age 30–49	Age 50+
Non-Hispanic White Alone	Males	**1.5**	**−2.1**	**0.6**
	Females	**1.1**	**0.7**	**2.2**
Black Alone or in Combination	Males	**−5.9**	**−10.0**	**−2.8**
	Females	0.4	0.2	**3.1**
American Indian/Alaskan Native Alone or in Combination	Males	−1.9	**3.1**	1.5
	Females	**−2.6**	0.4	**4.6**
Asian Alone or in Combination	Males	−1.2	**−2.2**	1.1
	Females	0.8	0.0	1.0
Native Hawaiian and Pacific Islanders Alone or in Combination	Males	**−8.0**	2.0	−0.4
	Females	1.9	−1.2	−1.4
Hispanic	Males	**−5.2**	**−5.1**	0.7
	Females	**−2.1**	−0.2	**3.6**

Source U.S. Census Bureau (2012a), Table C
A negative sign reflects a net undercount. The signs here are reversed from the source report in order to keep directionality consistent within this publication
Figures in **BOLD** are statistically significantly different than zero

In the Black Alone population, the biggest difference in coverage rates between males and females is for the population in their 30s and 40s. The biggest single year difference is age 29 where there is an 8.6% net undercount for males and a 0.7% overcount for females (data not shown here).

Table 6.3 shows Census coverage for males and females by age in six different race/Hispanic groups based on DSE. In almost every comparison in Table 6.3, females have better coverage rates than males. In a few cases the differences are extreme. For example, for Blacks Alone or in Combination age 30–49, the net undercount rate for males was 10% compared to a slight net overcount (0.2%) for females in this age range. This comparison highlights the very precarious position of Black males in their 30s and 40s in terms of being missed in the Census.

Table 6.3 also highlights the difference sex makes in the 18–49-year-old category compared to age 50 and over. For the population age 50 and over most groups experience a net overcount and for those groups that did experience a net undercount it is lower than that for age 18–49.

Table 6.4   2010 Census net undercount rates for males and females by age, sex, and tenure

	Population living in owner-occupied housing units		Population living in renter- occupied housing units	
	Males	Females	Males	Females
Age 18–29	**1.7**	**1.6**	**−3.9**	**−0.9**
Age 30–49	**−2.6**	**0.5**	**−5.5**	0.2
Age 50+	0.2	**2.0**	0.7	**3.5**

Source U.S. Census Bureau (2012a), Table A
A negative sign reflects a net undercount. The signs here are reversed from the source report in order to keep directionality consistent within this publication
Figures in **BOLD** are statistically significantly different from zero

Table 6.5   2010 census omission rates for male and females by age, sex, and tenure

Omissions rates	Population living in owner-occupied housing units		Population living in renter- occupied housing units	
	Males	Females	Males	Females
Age 18–29	5.9	5.7	12.5	9.2
Age 30–49	6.4	2.9	12.3	6.3
Age 50+	3.6	1.7	6.9	3.2

Source U.S. Census Bureau (2012a), Table A

6.4   Net Undercount and Omissions Rates for Males and Females by Age and Tenure

Table 6.4 provides net undercount rates for males and females by age and tenure. For the population living in owner-occupied housing units, females are covered better than males and the same is true for the population living in rental occupied housing units. The gap between males and females is smaller for the population living in owner-occupied housing than the population living in rental housing. The group with the highest net undercount rate is male renters age 30–49 where there is a net undercount rate of 5.5%. This group combines three characteristics (male, renter, and age 30–49) that are associated with poor coverage in the Census.

As stated earlier, DSE is the only method that provides omissions rates. Table 6.5 shows omissions rates by age, sex, and tenure from the 2010 Census. The age/sex patterns for omissions are like previously seen patterns for net undercounts. Males have higher omissions rates than females in every age group. The omissions rates for male renters age 18–29 are the highest at 12.5%.

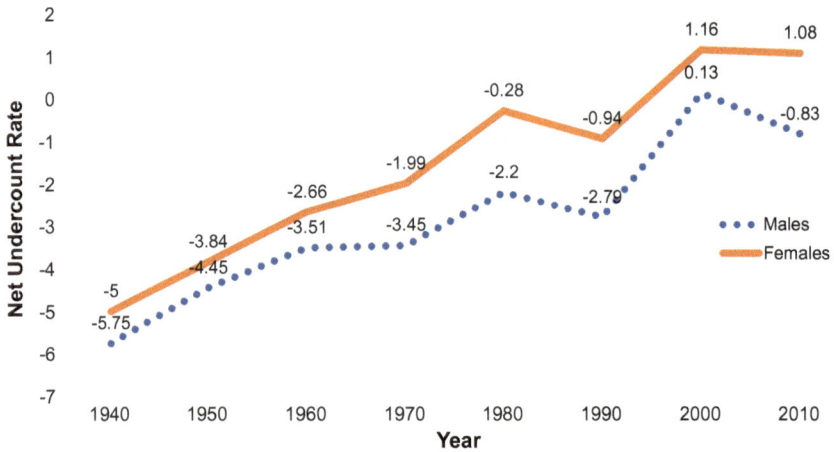

Fig. 6.2 Census net undecount rates for males and females: 1940–2010. *Source* 2010 Data from May 2012 DA Release. 1940–2000 from U.S. Census Bureau (2001), Table 4

6.5 Differential Census Coverage by Sex Over Time

Figure 6.2 shows net undercount rates for males and females from 1940 to 2010 based on Demographic Analysis. There are three key points shown in Fig. 6.2.

First, the coverage of females has been better than the coverage of males in every Census since 1940. Second, the coverage of both males and females has improved consistently since the 1940 Census. Both males and females experienced high net undercounts in the 1940 Census (5% or more), but in the 2010 Census there was 1.1% overcount for females compared to a 0.8% undercount for males.

Third, the difference in Census coverage between males and females has widened somewhat since 1940. In 1940, the difference between the coverage rate of males and females was less than one percentage point but in 2010 the difference was nearly two percentage points.

6.6 Sex and Sexual Orientation

I would be remiss if I did not say a word about sexual identity and sexual orientation in the context of the Census. The "sex" question was introduced in the first Census in 1790 and has been used ever since without generating much controversy. Yet as social norms about gender identity and sexual orientation have changed in recent decades, advocates began to press the Census Bureau for expanded questions and response options on sexual orientation and gender identity. The U.S. General Accountability

Office (2018, p. 5) identified "Lesbian gay bisexual transgender queer/questioning persons" as being hard-to-count.

In 1990, advocates ran the first national campaign to press the Bureau to report data on same-sex couples (National LGBTQ Task Force 2017). In 2005, the Census Bureau responded by beginning to test and report on data on same-sex couples in response to the "relationship" question (U.S. Census Bureau 2018). Starting in 2013, all Census tables that have a line for "married couples" include same-sex couples.

As we approach the 2020 Census, including measures of sexual orientation and gender identity on the 2020 Census questionnaire has become a widely discussed topic (Wang 2017). Modifications to the "relationship" question on the 2020 Census questionnaire will allow respondents to explicitly identify themselves as the married or unmarried same-sex partner of the householder for the first time. Other questions on gender identity and sexual orientation remain in the testing phase for possible inclusion on future Censuses or American Community Survey questionnaires. For more information about his topic see www.queertheCensus.org.

6.7 Summary

Generally, males have higher net undercount rates and higher omissions rates than females. There is virtually no gap between males and females for children (under age 18) or for the population over age 70. For adults in the 20s, 30s and 40s, males typically have a much a higher net undercount rates and higher omissions rates that females.

The male-female gap in coverage is larger for the population living in rental housing units than those living in owner-occupied units. The coverage gap between males and females is particularly large in the Black population.

The net undercount rates of males and females have both improved since the 1940 Census, but the gap between males and females has grown a little wider in recent decades.

References

Bhaskar, R., Scopilliti, M., Hollman, F., & Armstrong, D. (2010). *Plans for producing estimates of net international migration for the 2010 demographic analysis estimates*. Census Bureau Working Paper No. 90.

Devine, J., Sink, L., DeSalvo, B., & Cortes R. (2010). *The use of vital statistics in the 2010 demographic analysis estimates*. Census Bureau Working Paper No. 88.

Martin, E. (2007). Strength of attachment: Survey coverage of people with tenuous ties to residences. *Demography, 44*(2), 427–440.

National LGBTQ Task Force. (2017). *LGBTQ census advocacy, 1990–2017*. Available at: http://www.thetaskforce.org/wp-content/uploads/2017/05/LGBTQ-Census-Advocacy.pdf.

O'Hare, W. P., Robinson, J. G., West, K., & Mule, T. (2016). Comparing the U.S. decennial census coverage estimates for children from the demographic analysis and coverage measurement surveys. *Population Research and Policy Review*, 35(5), 685–704.

Robinson, J. G., (2010). *Coverage of population in census 2000 based on demographic analysis: The history behind the numbers*. U.S. Census Bureau, Working Paper No. 91.

Simpson, S., & Middleton, E. (1997). *Who is missed by a national census? A review of empirical results from Australia, Britain, Canada, and the USA*. UK: the Cathie Marsh Centre for Census and Survey Research, University of Manchester.

U.S. Census Bureau. (2001) *Accuracy and coverage evaluation: Demographic analysis results*. DSSD 2000 Procedures and Operation Memorandum Series B-4* J. G. Robinson.

U.S. Census Bureau. (2012a). *2010 Components of census coverage for race groups and Hispanic origin by age, sex, and tenure in the United States*. DSSD 2010 Census coverage Measurement Memorandum Series #2010-E-51, Table D. Washington, DC: U.S. Census Bureau.

U.S. Census Bureau. (2012b). *2010 Census coverage measurement estimation report: Summary of estimates of coverage for persons in the United States*. DSSD 2010 Census Coverage Measurement Memorandum Series #2010-G-01. Washington, DC: U.S. Census Bureau.

U.S. Census Bureau. (2018). *Frequently asked questions about same-sex couple households*. Available at: https://www2.Census.gov/topics/families/same-sex-couples/faq/sscplfactsheet-final.pdf.

U.S. General Accountability Office. (2018). *2020 Census: Actions needed to address challenges to enumerate hard-to-count groups*. Report to Congressional Requesters, GAO-18-599, Washington DC.

Wang, H. L. (2017). *Collecting LGBT census data is "Essential" to Federal Agency, documents shows*. National Public Radio, June 20.

Open Access This chapter is licensed under the terms of the Creative Commons Attribution 4.0 International License (http://creativecommons.org/licenses/by/4.0/), which permits use, sharing, adaptation, distribution and reproduction in any medium or format, as long as you give appropriate credit to the original author(s) and the source, provide a link to the Creative Commons license and indicate if changes were made.

The images or other third party material in this chapter are included in the chapter's Creative Commons license, unless indicated otherwise in a credit line to the material. If material is not included in the chapter's Creative Commons license and your intended use is not permitted by statutory regulation or exceeds the permitted use, you will need to obtain permission directly from the copyright holder.

Chapter 7
Census Coverage of the Hispanic Population

Abstract The net undercount for Hispanics in the 2010 Census was relatively high at 1.5% compared to a net overcount of 0.8% for the Non-Hispanic White Alone population. The omissions rate for Hispanics (7.7%) was about twice the rate for Non-Hispanic White Alone (3.8%). Hispanic men age 18–49 had net undercount rates over 5% and omissions rates over 10% in the 2010 Census.

7.1 Introduction

Studying the recent history of Census coverage of Hispanics is important because they are the fastest growing major race/ethnic group in the U.S. and they have relatively high net undercount and omissions rates. The number of Hispanics in the U.S. grew from 50.5 million in 2010 to 57.5 million in 2016 (U.S. Census Bureau 2017c). Consequently, the Census coverage of Hispanics increasingly drives the overall Census coverage figures.

The Dual-Systems Estimates (DSE) method provides net undercount estimates and omissions rates for Hispanic by age-sex groups and tenure in the 2010 Census. Demographic Analysis (DA) only provides data by single year of age for Hispanics under age 20 because Hispanic were not recorded on birth and deaths certificates consistently until 1990. The term Hispanic is used here rather than Latino, because that is the term used most often in Census Bureau publications.

7.2 Net Undercount Rates of Hispanic Adults

Overall, there was a net undercount for Hispanics (1.5%) and a net overcount of Non-Hispanic White Alone (0.8%) resulting in a difference of 2.3 percentage points. This pattern is generally consistent across age and sex groups.

Table 7.1 shows net undercount rates from DSE for Hispanics and the Non-Hispanic White Alone population ages 10 and over from the 2010 Census broken down by age and sex. Data for the youngest age group (0–9) are not presented in

© The Author(s) 2019

W. P. O'Hare, *Differential Undercounts in the U.S. Census*,
SpringerBriefs in Population Studies, https://doi.org/10.1007/978-3-030-10973-8_7

Table 7.1 2010 Census coverage of Hispanics compared to Non-Hispanic White Alone by age and sex

	Percent undercount	
	Hispanic	Non-Hispanic White Alone
Total population	**−1.5**	**0.8**
10–17	−0.1	**1.8**
18–29 male	**−5.2**	**1.5**
18–29 female	**−2.1**	**1.1**
30–49 male	**−5.1**	**−2.1**
30–49 female	−0.2	**0.7**
50+ male	0.7	**0.6**
50+ female	**3.6**	**2.2**

Source U.S. Census Bureau (2012c), Table C
A negative sign reflects an undercount. The signs here are reversed from the source report in order to keep directionality consistent within this publication
Figures in **BOLD** are statistically significantly different from zero

Table 7.1 because there is strong evidence that the coverage estimates of young children from DSE in 2010 are problematic (O'Hare et al. 2016). For information on the Census coverage of young Hispanic children in the 2010 Census see the analysis of DA data later in this Chapter.

Every age/sex group of Hispanics except those over age 50, had a net undercount while only one group of Non-Hispanic White Alone population (males age 30–49) had a net undercount. The net undercount rate for Hispanic males age 18–29 was 5.2% and the net undercount rate for Hispanic males 30–49 was 5.1% and both rates are statistically significantly different than zero.

To focus on differential coverage, the net undercounts for Hispanics are compared to the Census coverage rates for Non-Hispanic White Alone population in the same age/sex category. The biggest coverage differential in Table 7.1 is among males age 18–29. The net undercount rate for Hispanic males age 18–29 is 5.2% compared to a net overcount of 1.5% for Non-Hispanic White Alone. This results in a difference of 6.7 percentage points. This percentage point differential is more than twice as high as any other age/sex group examined in Table 7.1 and underscores the special problems associated with trying to enumerate Hispanic males in their 20s.

The differential Census coverage for males age 18–29 may reflect differences in socio-economic status. Overcounts in this age group are often attributed to young people being counted in the college dormitory (or off campus residence) as well as their parents' home (see Chap. 5). Census undercounts are often attributed to unstable living arrangements or lack of attachment to a housing unit (Martin 1999, 2007). The undercount differential reported above suggests that young adult Hispanic males may be more likely to be experiencing unstable living situations while young adult Non-Hispanic White Alone males may be more likely to be in college.

The net undercount rates for Hispanic males age 30–49 was also relatively high (5.1%) but the differential between Hispanics and Non-Hispanic Whites Alone is not as high as for males age 18–29 because there is a statistically significant net undercount for Non-Hispanic White Alone males in this age group (2.1%) as well, leaving a net undercount differential of only 3.0 percentage points.

It is possible that the net undercount rate for Hispanic males age 18–49 is even higher than what was reported by the Census Bureau because correlation bias among Hispanic men is not accounted for in the DSE analysis. Correlation bias means the types of people missed in the Census are also likely to be missed in the Post-Enumeration Survey that is the basis for the DSE estimates (U.S. Census Bureau 2012a). Correlation bias leads to an underestimate of net undercount rates. The Census Bureau made an adjustment to the DSE estimates for Black adult males, based on the sex ratio shown in DA (U.S. Census Bureau 2012a). No adjustment was made for Hispanic adult males because the data needed for such an adjustment (DA sex ratios) were not available for Hispanics.

Another group that is noteworthy in Table 7.1 is Hispanic females age 50 or older. There is a large statistically significant net overcount of this group (3.6%). This is the only age/sex group of Hispanics to experience a statistically significant net overcount. There is also a large statistically significant overcount for Non-Hispanic White Alone females in this age group (2.2%) so the gap between Hispanics and Non-Hispanic White Alone is relatively small. It is possible the high net overcount of Hispanic females age 50 or older is due to an under-reporting of births more than a half-century ago and/or undetected in-migration.

7.3 Omissions Rates for Hispanics

Recall that the Census net undercount rate is a balance between people omitted and those included erroneously (mostly double counted). The omissions rate captures the share of a group missed in the Census. DSE is the only method that shows omissions rates.

In many ways the omissions rate is a more meaningful statistic than the net under-count rate because in the net undercount calculation omissions can be cancelled out by erroneous inclusions or double counting. A net undercount of zero could be the result of no one being missed and no one double counted, or for example, 10% missed, and 10% double counted.

Table 7.2 shows omissions rates for Hispanics and Non-Hispanic White Alone by age and sex. The overall omissions rate for Hispanics (7.7%) is twice that of the Non-Hispanic White Alone population (3.8%). To a large extent the omissions rates reflect the same age/sex pattern as the net undercount rates. In every age/sex group examined here, the omissions rate for Hispanics is higher than that of Non-Hispanic White Alone. Like the results for net undercount rates, Hispanic males age 18–49 had the highest omissions rates. The omissions rate for Hispanic males age 18–29 was 12.4% and the omissions rate for Hispanic males age 30–49 was 10.9%.

Table 7.2 2010 Census omissions rates for Hispanics and Non-Hispanic Whites Alone by age and sex

	Percent omissions	
	Hispanic	Non-Hispanic White Alone
Total population	7.7	3.8
10–17	5.9	3.1
18–29 male	12.4	6.6
18–29 female	9.6	6.2
30–49 male	10.9	6.2
30–49 female	5.8	3.0
50+ male	5.5	3.5
50+ female	2.5	1.7

Source U.S. Census Bureau (2012c), Table C

Hispanic females age 18–29, also had a relatively high omissions rate of 9.6% but the omissions rates for Hispanic females age 30–49 (5.8%) was lower than the overall omissions rate for Hispanics.

7.4   Differences in Census Coverage by Tenure

Table 7.3 shows net undercount rates and omissions rates from the 2010 Census DSE analysis for the populations living in owner-occupied housing units and renter-occupied housing units among Hispanics and the Non-Hispanic White Alone population.

Two patterns are clear. First, Non-Hispanic White Alone population was covered better than Hispanics for the populations living in both owner-occupied housing units and rental housing units. Among the population living in owner-occupied units, there was a larger net overcount for Non-Hispanic Whites Alone (0.8%) than for Hispanics (0.3%) and the rate for Non-Hispanic Whites Alone was statistically significantly different than zero, while the rate for Hispanics was not. Among those living in renter-occupied housing units there was a statistically significant net undercount for Hispanics (3.3%) compared to a net overcount (0.9%) for Non-Hispanic White Alone. Omissions rates for Hispanics are higher than for Non-Hispanic White Alone for renters and owners.

Second, the gap between census coverage of Hispanics and Non-Hispanic Whites Alone is bigger for the population living in renter-occupied housing units than for the population living in owner-occupied housing units.

For some categories of Hispanic renters, the omissions rates are particularly high. The compound impact of Hispanic Origin Status, age, and tenure can be seen by looking at the omissions rates of Hispanic male renters age 18–29 where about one-

Table 7.3 2010 Census net undercount rates and omissions rates for Hispanics and Non-Hispanic White Alone by tenure

		Hispanic origin	Non-Hispanic White Alone
Percent undercount	Population living in owner-occupied housing units	0.3	**0.8**
	Population living in renter-occupied housing units	**−3.3**	0.9
Percent omissions	Population living in owner-occupied housing units	5.0	3.0
	Population living in renter-occupied housing units	10.4	6.4

Source U.S. Census Bureau (2012c), Table B
A negative sign reflects a net undercount. The signs here are reversed from the source report in order to keep directionality consistent within this publication
Figures in **BOLD** are statistically significantly different from zero

sixth (16.1%) were missed in the 2010 Census and nearly one-seventh (14.1%) of Hispanic males age 30–49 living in rental housing units were missed in the 2010 Census (U.S. Census Bureau 2012b, Table C).

7.5 Census Coverage of Hispanic Children Age 0–19

In the 2010 Census, DA undercount estimates for Hispanics are limited to those under age 20 because it has only been since 1990 that Hispanics were systematically identified in birth and death certificates across all the states. In the 2010 Census cycle, the Census Bureau first produced a series of five DA estimates based on differing assumptions about births, deaths, and net international migration. The Census Bureau produced five different series to reflect some uncertainty in the DA estimates. The estimates were released in early December of 2010, to mitigate any perception that the actual Census count (released in late December) might influence the DA estimates. Subsequently, the Census Bureau updated the middle series DA estimate in 2012 but did not include Hispanics in the update because the updated data were primarily for use to develop sex ratios for Black adults to be used in the DSE estimates.

Figure 7.1 shows net undercount rates for Hispanic and Non-Hispanics by single year of age from 0 to 19, based on data from the Middle Series December 2010 DA release. Results for Hispanics are compared to those for Non-Hispanics because there are no data for Whites or Non-Hispanic Whites in the DA results for 2010. It

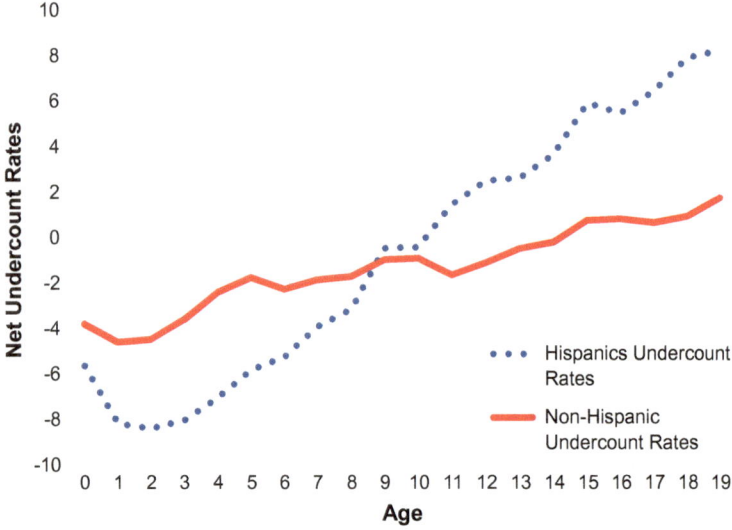

Fig. 7.1 2010 Census net undercount rates by single year of age for Hispanics and Non-Hispanics: Age 0–19. *Source* U.S. Census Bureau, December 2010 DA Release

should be noted that Non-Hispanics include Blacks and American Indians and these population have above average net undercount rates (see Chap. 4).

When the population ages 0–19 are examined collectively, Hispanics and Non-Hispanics show very similar net undercount rates. The net undercount rate for Hispanics age 0–19 is 1.2% and for Non-Hispanics it is 1.3%. However, these averages mask some important age differences among Hispanics and Non-Hispanics. At the youngest ages, the net undercount rate for Hispanics is much higher than that for Non-Hispanics. Based on the DA data released in 2010, there was a net undercount of 7.5% for Hispanics age 0–4 compared to a 3.6% net undercount for Non-Hispanics age 0–4 (O'Hare 2015). O'Hare (2015) also reports the net undercount rate for Hispanic children age 0–4 was 7.7% in the 2000 Census. But for ages 10–19, Hispanics had a higher net overcount than Non-Hispanics. For age 10–19, there was a net overcount of 4.3% for Hispanics and 0.1% for Non-Hispanics.

There are a couple of methodological notes on this topic that readers should be aware of. First, there is reason to believe that the DA estimates for the teenage Hispanics may be problematic because of net immigration assumptions for teenagers that did not take into consideration the economic downtown during the 2008 to 2010 period, which may have dampened in-migration of young Hispanics. But this probably had a small impact on the estimates.

Second, the data for age zero (those less than one year old) are problematic for both Hispanics and Non-Hispanics because Hogan and Griffin (2017) found many infants born after April 1, 2010, were included in the 2010 Census count erroneously. So, the net undercount rates for age 0 is actually higher than what is shown here.

Third, it is possible the high net undercount for the youngest Hispanics is partly due to assumptions about births to Hispanics in the 2008–2010 period that were too high. When the DA estimates were first issued in December 2010, the Census Bureau had to make assumptions about births and deaths in 2008, 2009 and the first quarter of 2010 because data from National Center for Health Statistics were not yet available. Recently, Jensen et al. (2016) provided updated estimates for young Hispanics in 2010 based on observed births and deaths in 2008–2010. The update suggests that the initial estimated net undercount for young Hispanics was a little high. The original net undercount estimate was 7.5% and the revised estimate was 6.5%.

Fourth, recently, the U.S. Census Bureau discovered data from the 2010 Mexican Census that shed light on net immigration estimates for young U.S.-born Hispanic children in 2010 (Jensen et al. 2018). The new data suggests earlier undercount estimates for young Hispanics were too high.

Collectively the new data generally show a lower net undercount rate for young Hispanics children than the original DA estimates, but the new net undercount estimates for young Hispanics are still very high compared to other demographic groups.

7.6 Bilingual Questionnaires

In one respect, Hispanics are different from other groups because bilingual (Spanish/English) Census questionnaires were sent to households in many neighborhoods with large Spanish-speaking populations in the 2010 Census (U.S. Census Bureau 2011). The 2010 Census was the first one to use a bilingual English/Spanish questionnaire. Almost half (46%) of all Hispanics in the U.S. resided in areas that received bilingual questionnaires in the 2010 Census.

Table 7.4 shows the net undercount and omissions rates for Hispanics and Non-Hispanics in Bilingual Mailing Areas compared to other areas. In both types of areas, Hispanics had a statistically significant net undercount. The Hispanic net undercount in Bilingual Mailing Areas (1.3%) is lower than that in other areas (1.7%), but the difference is small. The omissions rate for Hispanics in both types of areas are almost identical (7.9% in bilingual areas compared to 7.6% in other areas).

After evaluating the 2010 Census experience with bilingual questionnaires, the Census Bureau (2011, p. v) concluded, "Further results suggest that the bilingual questionnaire provides substantial benefit to the areas that were targeted…"

7.7 Hispanic Trend Data from 1990 to 2010

Table 7.5 shows estimated net undercount rates for 1990, 2000 and 2010 Censuses for Hispanics and Non-Hispanic Whites Alone based on the Census Bureau's DSE

Table 7.4   2010 Census undercount and omissions rates for Hispanics and Non-Hispanics inside and outside bilingual mailing areas

	Census count (in 1000s)	Percent undercount	Percent omissions
U.S. total	3,00,703	0.0	5.3
Bilingual mailing area	35,204	**−0.8**	7.3
Hispanic	22,498	**−1.3**	7.9
Non-Hispanic	12,706	0.2	6.0
Balance of U.S.	2,65,499	0.1	5.1
Hispanic	27,082	**−1.7**	7.6
Non-Hispanic	2,38,418	**0.3**	4.8

Source U.S. Census Bureau (2012b), Table 16
A negative sign reflects an undercount. The signs here are reversed from the source report in order to keep directionality consistent within this publication
Percent undercount figures in **BOLD** are statistically significantly different from zero

Table 7.5   Estimates of net undercount rates of Hispanics and Non-Hispanic White Alone populations: 1990, 2000 and 2010

	2010	2000	1990
	Percent undercount	Percent undercount	Percent undercount
U.S. total	0.0	**0.5**	**−1.6**
Non-Hispanic White Alone	**0.8**	**1.1**	**−0.7**
Hispanic	**−1.5**	−0.7	**−5.0**

Source U.S. Census Bureau (2012b), Table 7
A negative sign reflects an undercount. The signs here are reversed from the source report in order to keep directionality consistent within this publication
Figures in **BOLD** are statistically significantly different from zero

method. In terms of trends over time, the coverage of Non-Hispanic White Alone is consistently better than that of Hispanics, but the magnitude of the difference varies over time.

For Hispanics, there was a relatively high net undercount rate in 1990 (5.0%) but in the 2000 Census the net undercount rate for Hispanics (0.7%) was so low that it was not statistically significantly different than zero. In the 2010 Census, however, there was an estimated net undercount of 1.5% for Hispanics which was statistically significant. The undercount rate for Non-Hispanic Whites Alone went from a net undercount of 0.7% in 1990, to a net overcount of 1.1% in 2000, to a net overcount of 0.8% in 2010.

Looking at net undercount rates for both Hispanics and Non-Hispanic Whites from 1990 to 2010 there doesn't seem to be any consistent trend over the 1990–2010 period. The difference between the Census coverage rates of Hispanics and Non-Hispanic White Alone decreased from 1990 to 2000 but increased from 2000 to 2010.

7.8 Census Coverage of Hispanic Subgroups

All the Census coverage estimates on Hispanics produced by the Census Bureau treat Hispanics as a single group. But the Hispanic population is far from homogeneous. There are several distinctions that should to be made in the Hispanic population with respect to Census coverage.

First, there are several subgroups of Hispanics such as Mexicans, Puerto Ricans, Central and South Americans, and Cubans. The 2016 American Community Survey indicates there are 36.3 million Mexicans, 5.5 million Puerto Ricans, 5.3 million Central Americans, 3.5 million South Americans, and 2.2 million Cubans in the U.S. Hispanic population. The social, economic, and cultural differences among these subgroups of Hispanics suggest that they are likely to have different levels of Census coverage (National Research Council 2006; O'Hare 2017). There are no direct estimates of Census coverage rates for subgroups of Hispanics produced by the Census Bureau, but some researchers have produced estimates for some subgroups. Using multiple methods, one group of researchers, (Van Hook et al. 2014, p. 699) concluded, "Additionally, we find evidence that U.S. Census and ACS data miss substantial numbers of children of Mexican immigrants, as well as people who are most likely to be unauthorized: namely working-aged Mexican immigrants (ages 15–64) especially males."

Second, many Hispanics speak English well, but some do not. For those who do not speak English well, participating in the Census may be more difficult. About 73% of Hispanics speak a language other than English at home and 31% speak English less than "very well" which is typically the threshold used to determine limited English proficiency (The Leadership Conference Education Fund 2017).

Third, a large share of the Hispanic population are recent immigrants. Kissam (2017) provides ample evidence that recent Mexican immigrants in California had a high likelihood of being missed in the Census. Jensen et al. (2015) show that recent immigrants are more likely to be missed in the Census Bureau's American Community Survey. Many of the recent Hispanic immigrants fall into the category of cultural and linguistic minorities which Harkness et al. (2014) argue is a hard-to-count group in the Census.

Fourth, a substantial fraction of Hispanic immigrants is undocumented. Evidence suggests the undocumented population have high net undercount rates (Warren and Warren 2013; Van Hook et al. 2014). For example, Warren and Warren (2013, p. 307) estimate a net undercount rate of 10% in the 2000 Census for undocumented immigrants who entered the U.S in the 1990s. Also, Van Hook et al. (2014, p. 720) state, "Age and sex patterns further suggest that coverage error among unauthorized Mexican immigrants is probably higher than that for the entire Mexican-born population." It should also be noted that census coverage of some Hispanic subgroups has been addressed in qualitative studies (Romero 1992; Dominguez and Mahler 1993; Mahler 1993).

It is also important to note that Censuses in some other countries are not always viewed positively and immigrants may bring their negative views about census-taking with them when they move to the U.S.

The last-minute addition of a question on citizenship to the 2020 Census is likely to have big implications for the count of Hispanics in the 2020 Census (Ross 2018; Barabba and Flynn 2018; Meyers and Goerman 2018; U.S. Census Bureau 2017a, b). See Chap. 15 for more information on this issue.

7.9 Summary

The net undercount of Hispanics in the 2010 Census was 1.5% compared to a net overcount of 0.8% for the Non-Hispanic White Alone population. Other findings include:

- Hispanic Males age 18–49 had the highest net undercount rate and highest omissions rate of any age/sex group of Hispanics.
- Hispanic children age 0–4 had a very high net undercount rate (7.5% based on December 2010 DA).
- Hispanic females over age 50 had a high net overcount rate.
- The trend in the net undercount of Hispanics from 1990 to 2010 is complicated. The net undercount rate decreased between 1990 and 2000 but increased between 2000 and 2010.
- The Hispanic population is very diverse with respect to characteristics linked to being difficult to enumerate.

References

Barabba, V., & Flynn, K. H. (2018). Ensure everyone is counted. *U.S. News*, January 30, 2018.

Dominguez, B., & Mahler, S. (1993). *Alternative enumeration of undocumented Mexicans in South Bronx*. Prepared under Joint Statistical Agreement 89-46 with Columbia University, Bureau of the Census, Washington DC.

Harkness, J., Stange, M., Cibelli, K. L., Mohler, P., & Pennell, B. (2014). Surveying cultural and linguistic minorities. In R. Tourangeau, B. Edwards, T. P. Johnson, K. M. Wolter, & N. Bates (Eds.), *Hard-to-survey populations* (pp. 245–269). Cambridge, England: Cambridge University Press.

Hogan, H., & Griffin, D. (2017). *Net census coverage of very young children*. Presentation at the Annual Conference of the Southern Demographic Association, October 26, Morgantown WV.

Jensen, E. B., Benetsky, M., & Garrow, S. (2016). *Updating the 2010 demographic analysis estimates of the Hispanic population*. Presentation at the 2016 Southern Demographic Association Conference, Athens GA.

Jensen, E., Benetsky, M., & Knapp, A. (2018). *A sensitivity analysis of the net undercount for young Hispanic children in the 2010 census*. Poster presented at the Population Association of America annual Conference, Denver, CO, April.

Jensen, E. B., Bhaskar, R., & Scopilliti, M. (2015). *Demographic analysis 2010: Estimates of coverage of the foreign-born population in the American Community Survey.* U.S. Census Bureau. Population Division U.S. Census Bureau Working Paper No. 103, June.

Kissam, E. (2017). Differential undercount of Mexican immigrant families in the U.S. census. *Statistical Journal of the International Association of Official Statistics.*

Mahler, S. (1993). *Alternative enumeration of undocumented Salvadorans on Long Island.* Prepared under Joint Statistical Agreement 89-46 with Columbia University, Bureau of the Census, Washington DC.

Martin, E. (1999). Who knows who lives here? Within-household disagreements as a source of survey coverage error. *Public Opinion Quarterly, 63,* 220–236.

Martin, E. (2007). Strength of attachment: Survey coverage of people with tenuous ties to residences. *Demography, 44*(2), 437–440.

Meyers, M., & Goerman, P. (2018). *Respondents confidentiality concerns in multilingual pretesting studies and possible effects on response rates and data quality for the 2020 census.* Paper delivered at the annual Conference of the American Association of Public Opinion Research, Denver CO, May 16–19, 2018.

National Research Council. (2006). *Hispanics and the future of America.* In M. Tienda & F. Michell (Eds.). Washington, DC: The National Academies Press.

O'Hare, W. P. (2015). *The undercount of young children in the U.S. Decennial Census.* Springer Publishers.

O'Hare, W. P. (2017). *Counting all Californians in the 2020 census.* The Census Project. https://Censusproject.files.wordpress.com/2017/10/calif-report-10-16-2017_format-final.pdf.

O'Hare, W. P., Robinson, J. G., West, K., & Mule, T. (2016). Comparing the U.S. decennial census coverage estimates for children from the demographic analysis and coverage measurement surveys. Forthcoming in *Population Research and Policy Review* (https://doi.org/10.1007/s11113-016-9397-x).

Romero, M. (1992). *Ethnographic evaluation of behavioral causes of census undercount of undocumented immigratns and Salvadorans in the Mission District of San Francisco.* U.S. Census Bureau. https://www.Census.gov/srd/papers/pdf/ev92-18.pdf.

Ross, W. (2018). *Reinstatement of a citizenship question on the 2020 decennial census questionnaire.* Memorandum from Secretary of Commerce Wilber Ross to Undersecretary of Commerce Karen Dunn Kelley, May 26, 2018.

The Leadership Conference Education Fund. (2017). *Factsheet: Will you count? Latinos in the 2020 census.*

U.S. Census Bureau. (2011). *2010 Census: Bilingual questionnaire assessment report.* 2010 Census Planning Memorandum Series, No. 156. Washington, DC: U.S. Census Bureau.

U.S. Census Bureau. (2012a). *2010 census coverage measurement estimation report: Adjustment for correlation bias.* DSSD 2010 Census Coverage Measurement Memorandum Series #2010-G-11. Washington, DC: U.S. Census Bureau.

U.S. Census Bureau. (2012b). *2010 Census coverage measurement estimation report: Summary of estimates of coverage for persons in the United States.* DSSD 2010 Census Coverage Measurement Memorandum Series #2010-G-01. Washington, DC: U.S. Census Bureau.

U.S. Census Bureau. (2012c). *2010 Components of census coverage for race groups and Hispanic origin by age, sex, and tenure in the United States.* DSSD 2010 Census Coverage Measurement Memorandum Series #2010-E-51. Washington, DC: U.S. Census Bureau.

U.S. Census Bureau. (2017a). *Respondent confidentiality concerns.* Memorandum for Associate Directorate for Research and Methodology (ADRM) From Center for Survey Measurement (SCM), September 20, 2017.

U.S. Census Bureau. (2017b). *Respondent confidentiality concerns and possible effects on response rates and data quality for the 2020 census.* Presentation by Mikelyn Meyers at the Census Bureau National Advisory Committee Meeting, November 2, 2017.

U.S. Census Bureau. (2017c) Population estimates downloaded from on February 4, 2018. https://factfinder.Census.gov/faces/tableservices/jsf/pages/productview.xhtml?src=bkmk.

Van Hook, J., Bean, F., Bachmeier, J. D., & Tucker, C. (2014). Recent trends in coverage of the Mexican-born population of the United States: Results from applying multiple methods across time. *Demography, 51*(2), 699–726.

Warren, R., & Warren, J. R. (2013). Unauthorized immigration to the United States: Annual estimates and components of changes, by state, 1990 to 2010. *International Migration Review, 47*(2), 296–329.

Open Access This chapter is licensed under the terms of the Creative Commons Attribution 4.0 International License (http://creativecommons.org/licenses/by/4.0/), which permits use, sharing, adaptation, distribution and reproduction in any medium or format, as long as you give appropriate credit to the original author(s) and the source, provide a link to the Creative Commons license and indicate if changes were made.

The images or other third party material in this chapter are included in the chapter's Creative Commons license, unless indicated otherwise in a credit line to the material. If material is not included in the chapter's Creative Commons license and your intended use is not permitted by statutory regulation or exceeds the permitted use, you will need to obtain permission directly from the copyright holder.

Chapter 8
Census Coverage of the Black Population

Abstract In the 2010 Census, the Black population had the highest net undercount rate of any major race/Hispanic group. Based on the Demographic Analysis method there was a net undercount of 2.5% for the Black Alone population compared to a net overcount of 0.5% for Non-Blacks. Black males in their 20s, 30s, ands 40s had exceptionally high net undercount rates and high omissions rates. Historically, the Black population, and Black men in particular, experienced high net undercount rates in the Census. While the net undercount rates of Blacks have decreased over time, the differential net undercount between Blacks and Non-Blacks has improved little since 1940.

8.1 Introduction

Studying the U.S. Census coverage of the Black population is important because they have been undercounted in the Census for many decades and they are a large part of the U.S. population. Consequently, coverage of the Black population has a big impact on the overall accuracy of the Census.

In reporting Census data on the Black population, it is very important to be clear about how the group is defined. Starting in the 2000 Census, people were allowed to select more than one race in the Census questionnaire (U.S. Office of Management and Budget 1997) and race is often shown two different ways in Census Bureau reports. One category is the number of people who only select Black (referred to as Black Alone) and second category is all those in the first category plus those who select Black along with at least one other race (referred to as Black Alone or in Combination). In the 2016 American Community Survey conducted by the Census Bureau there were 40.9 million people who marked Black Alone and 45.1 million who marked Black Alone or Black and some additional race. The Black Alone or in Combination population comprised about 14% of the U.S. total population in 2016. Black Alone or in Combination is the most inclusive definition and its use here is consistent with advice of the U.S. Office of Management and Budget (2001) about using race classifications. Most of the data presented in this Chapter are based on the

© The Author(s) 2019   83
W. P. O'Hare, *Differential Undercounts in the U.S. Census*,
SpringerBriefs in Population Studies, https://doi.org/10.1007/978-3-030-10973-8_8

Black Alone or in Combination definition. The term Black is used instead of African-American because it is the label used most often in Census Bureau publications.

8.2 Census Coverage of the Black Population by Age and Sex

Both methods for measuring Census accuracy (Demographic Analysis—DA and Dual-Systems Estimates—DSE) show high net undercounts for the Black population. In the 2010 Census, the net undercount rate for the Black Alone population was 2.5% based on Demographic Analysis (DA) while for Non-Blacks there was a 0.5% net overcount (Velkoff 2011). Based on Dual Systems Estimates (DSE) results, the 2010 Census shows a net undercount for the Black Alone or in Combination population of 2.1% compared to a net overcount of 0.8% for the Non-Hispanic White Alone population (U.S. Census Bureau 2012).

The overall net undercount rate for the Black population masks large differences by age and sex. For example, there was a net undercount of 4.6% for Black males compared to a 0.1% net undercount for Black females in the 2010 Census based on DA.

The Black adult male population is particularly vulnerable to net undercounts in the Census. Figure 8.1 shows net undercount rates in the 2010 Census for Black Alone males, Black Alone females, Non-Black Alone males, and Non-Black Alone females by five-year age groups based on DA. Figure 8.1 shows that Black Alone males stand out from the other groups in terms of having high net undercounts from their 20s to their 70s.

Early in life and late in life there are not big differences in the coverage rates of Black males and the other three race/sex groups shown in Fig. 8.1. However, starting in their early 20s into their 70s, the undercount differentials between Black Alone males and the other three groups are substantial. For example, at age 30 the difference in Census coverage rates between Black Alone males and Black Alone females is about 8 percentage points (net undercount rate of 8.3% for 30-year-old Black Alone males and 0% for 30-year-old Black Alone females).

This pattern is not new. High net undercount rates for Black men have been a persistent problem in the U.S. Census (Fein 1989; Hill 1975; Passel 1991; Robinson et al. 1990; Robinson 1997; Fay et al. 1988; West et al. 2014). The report on DA following the 1990 Census (U.S. Census Bureau 1991, p. 1) states,

> To summarize, two groups stand out as having relatively high levels of net undercount nationally after considering the possible range of uncertainty in the estimates: 1) Black children aged 0-9 and 2) Black men aged 20-64.

Much of the public and political interest regarding Census undercounts in general and Census undercounts of Black men in particularly can be traced back to publications and events in the late 1960s and early 1970s (Valentine and Valentine 1971; Heer 1968). One of the first events to publicize the high net undercount of the black

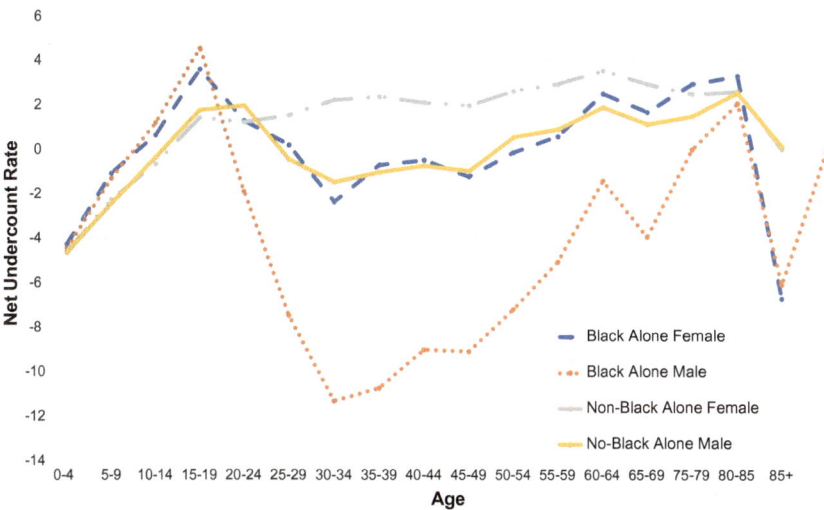

Fig. 8.1 2010 census net undercount rates by sex and Black/Non-Black by five-year age groups. *Source* US Census Bureau, May 2013

population was a 1968 conference organized by David Heer at the behest of Daniel Patrick Moynihan. According to Heer (1968, p. 2) "Moynihan asked me if I would plan a conference that would (1) publicize the fact that many Negroes had not been counted in the 1960 census, and (2) attempt to rouse national concern about the matter."

One crucial study was that of Hill (1975) who used a synthetic estimation technique and age/sex/race net undercount estimates from the 1970 Census to produce net undercount estimates for large cities. These estimates showed that many large cities had relatively high net undercounts largely because they housed a disproportionately large share of Black males. Hill's study documented something many big-city mayors suspected but had been unable to quantify or demonstrate previously.

To some extent the studies cited above, and other similar studies, launched a series of efforts by the Census Bureau to address differential undercounts (Choldin 1994). Those efforts involved advocating for adjustment of Census results based on net undercounts and better Census outreach in the Black community (see Chap. 13 for more details on this topic).

Table 8.1 shows net undercount rates by age and sex for Black Alone or in Combination and Non-Hispanic White Alone from DSE and they underscore the pattern shown in Fig. 8.1. Data for the youngest ages (0–9) are not presented in Table 8.1 because there is strong evidence that the coverage estimates of young children from DSE in 2010 are problematic (O'Hare et al. 2016).

There are three main points that can be drawn from the data in Table 8.1. First, the Black Alone or in Combination population were generally undercounted while the Non-Hispanic White Alone population was over counted. Second, Black Alone

Table 8.1 Net undercount rates for the Black Alone or in combination and Non-Hispanic White Alone populations in the 2010 census by age and sex

	Black Alone or in combination	Non-Hispanic White Alone
Total	**−2.1**	**0.8**
Age 10–17	−0.1	**1.8**
18–29 males	**−5.9**	**1.5**
18–29 semales	0.4	**1.1**
30–49 males	**−10.0**	**−2.1**
30–49 females	0.2	**0.7**
50+ males	**−2.8**	**0.6**
50+ females	**3.1**	**2.2**

Source US Census Bureau (2012). Table C
A negative sign reflects an undercount. The signs here are reversed from the source report in order to keep directionality consistent within this publication
Figures in **BOLD** are statistically significantly different from zero

or in Combination males were undercounted at a higher rate than Black Alone or in Combination females. Third, both Black Alone or in Combination and White Alone or in Combination females over age 50 have relatively high net overcount rates.

Data in Table 8.1 show net undercount rates for Black Alone or in Combination males in their 30s and 40 were among the highest recorded in the 2010 Census. Table 8.1 shows the net undercount rate for Blacks Alone or in Combination males age 18–29 was 5.9% and for Black Alone males age 30–49 was 10.0%. The results of the DA methodology are consistent with the DSE results. DA shows a net undercount rate for Black Alone males age 18–49 was 7.6%.

8.3 Census Coverage of Black Children

As noted in Chap. 5, the population age 0–4 had a higher net undercount rate than any other age group in the 2010 Census. There was a high net undercount of young Black children but the net undercount for Black Alone differs strikingly from Black Alone or in Combination. For the population age 0–4 the net undercount rate in the 2010 Census for Black Alone was 4.4% compared to 6.3% for Black Alone or in Combination (O'Hare 2015). Both rates are very high compared to other groups.

Much of the difference between Black Alone and Black Alone or in Combination net undercount rates probably stems from difficulties trying to make the race classifications based on birth certificates consistent with the race classifications based on self-reporting in the Census. Recall that race of a newborn on the birth certificate is assigned based on the race or the parents. Reconciling race from the Census and birth certificates has become more challenging since 2003 when the new birth certificates were introduced which allowed parents to mark more than one race. Complications with making race categories consistent between the Census and vital events were discussed in Chap. 3.

8.4 Census Omissions Rates for the Black Population

Recall that the net Census undercount rate is a balance between people omitted and those included erroneously (mostly double counted). The omissions rate captures the share of a group missed in the Census. DSE is the only method that shows omissions rates.

In many ways the omissions rate is a more meaningful statistic because in the net undercount calculation, omissions can be cancelled out by erroneous inclusions or double counting. A net undercount of zero could be the result of no one missed and no one double counted, or for example, ten% missed, and ten% double counted.

Table 8.2 shows the omissions rates for the Black Alone or in Combination population and the Non-Hispanic White Alone population by age and sex based on the DSE method. The omissions rate for the Black Alone or in Combination population in the 2010 Census was 9.3% compared to 3.8% for the Non-Hispanic White Alone population. But the omissions rate for the entire population masks substantial variations within age/sex subgroups of the population. Data for young children are not provided here because undercount estimates for young children from the DSE method are problematic (O'Hare et al. 2016).

The omissions rates shown in Table 8.2 follow a familiar pattern. In every demographic category examined in Table 8.2 the omissions rate for Blacks Alone or in Combination was higher than that of Non-Hispanic White Alone. In several cases the omissions rates for Blacks Alone or in Combination were more than twice as high as the rate for Non-Hispanic Whites Alone in the same category. The omissions

Table 8.2 Omissions rates for Black Alone or in combination population and Non-Hispanic White Alone in the 2010 census by age and sex

	Black Alone or in combination	Non-Hispanic White Alone
Total	9.3	3.8
Age 10–17	6.9	3.1
18–29 male	15.6	6.6
18–29 female	9.7	6.2
30–49 male	16.7	6.2
30–49 female	6.2	3.0
50+ male	9.2	3.5
50+ female	2.8	1.7

Source US Census Bureau (2012). Table C

rate for Blacks Alone or in Combination males age 18–29 was 15.6% and for age
30–49 it was 16.7%. Data in Table 8.2 reinforces the very precarious situation of
Black men age 18–49 in terms of the probability of being missed in the Census.

8.5 Net Coverage by Tenure

Table 8.3 shows net undercount rates and omissions rates for Black Alone or in Combination and Non-Hispanic White Alone from the 2010 Census by tenure. For both
the population living in owner-occupied housing units and the population living in
renter-occupied housing units, Blacks Alone or in Combination have net undercounts
while Non-Hispanic Whites Alone have net over counts. The difference in Census
coverage between Black Alone or in Combination and Non-Hispanic Whites Alone
is bigger among renters than among owners.

The omissions rates for Black Alone or in Combination renters is particularly high
at 11%. For some categories of Black Alone or in Combination renters the omissions
rates are extremely high. The compound impact of race, age, and tenure is reflected
in omissions rates for Black males age 30–49 living in rental housing units where
the omissions rate is nearly one-fifth (19.7%) and about one-sixth (16.9%) of Black
male renters age 18–29 were missed in the 2010 Census (U.S. Census Bureau 2012,
Table C).

Table 8.3 2010 Census net undercount rates and omissions rates for Black Alone or in combination
and Non-Hispanic White Alone by tenure

		Black Alone or in combination	Non-Hispanic White Alone
Percent undercount	Population living in owner-occupied housing units	**−0.9**	**0.8**
	Population living in renter-occupied housing units	**−3.0**	0.9
Percent omissions	Population living in owner-occupied housing units	7.2	3.0
	Population living in renter-occupied housing units	11	6.4

Source US Census Bureau (2012). Table B
A negative sign reflects a net undercount. The signs here are reversed from the source report in
order to keep directionality consistent within this publication
Figures in **BOLD** are statistically significantly different from zero

8.6 Census Coverage of the Black Population Over Time

Historically, DA estimates have only been available for two groups: Black and Non-Black. This restriction is due to the lack of race specificity and consistency for data collected on the birth and death certificates historically. The only group that has been identified relatively consistently over time is Blacks and the residual category has been Non-Black. It is important to recognize that the Black population estimates from DA have not been completely consistent over the past several decades (see Robinson 2010) but these inconsistencies are unlikely to have a major impact on trends over time.

In some ways the changes in the Census coverage rates for the Black population over the past 70 years is a "good news/bad news" story. The good news is that the net undercount rate for the Black population has decreased dramatically since 1940. The bad news is that the difference between Census coverage rates for the Black population and Non-Black population has changed very little over that time period.

Figure 8.2 shows the net undercount rate for the Black population has gone from 8.4% in 1940 to 2.5% in 2010. While this is a substantial improvement in Census coverage since 1940, the Black net undercount rate in 2010 is still the highest of any major racial/ethnic group. Moreover, despite this substantial improvement in the Census coverage of the Black population since 1940, the differential undercount between Blacks and Non-Blacks has not changed much. Figure 8.2 shows the differential undercount (Non-Black undercount rate minus the Black undercount rate) was 3.4% age points in 1940 and 3.0% age points in 2010.

However, the short-term story (1990–2010) is somewhat different than the long-term story (1940–2010). The net undercount rate of 5.5% for Blacks in the 1990 Census fell to 2.5% in the 2010 Census. The differential undercount fell from 4.4 points in 1990 to 3.0 in 2010 driven largely by improvements in the coverage of the Black population.

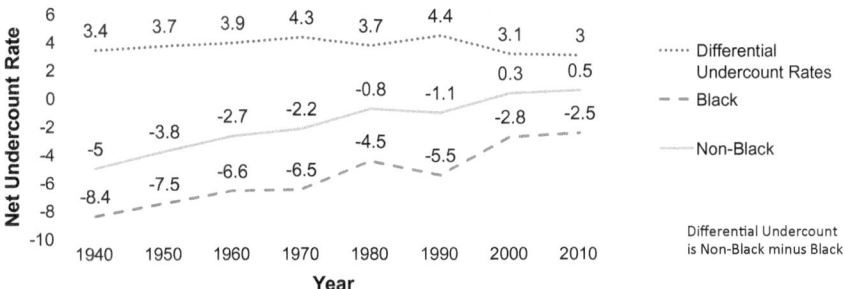

Fig. 8.2 Differential undercount rates for Blacks and Non-Blacks: 1940–2010. *Sources* Data for 1940–2000 from envisioning the 2020 Census, National Research Council, (2010) Table 2-1; Data for 2010 from Velkoff (2011) Table 2-1, p. 30

I am not aware of any commonly accepted explanation for the dramatic improvement in the Census coverage of the Black population between 1990 and 2010. It is possible the improvement is related to expanded outreach activities such as the Census Partnership Program and Paid Advertising started in the 2000 Census. It is also possible the large-scale incarceration of Black men resulted in a lower net undercount. Finally, changes between 1990 and 2010 in the way data on race have been collected may be related to the changes in net undercount of the Black population since 1990.

It should be noted that the composition of the Non-Black population has changed over time, particularly in the last few decades. Hispanics, another group with a relatively high net undercount, have become a larger part of the Non-Black population which confounds comparisons of Blacks and Non-Blacks over time.

8.7   Summary

The 2010 Census net undercount and omissions rates for the Black population were the highest of any major race or ethnic group. Black adult males have had particularly high net undercount and omissions rates. The relatively high net undercount of Blacks has been seen in every Census since 1940. Some demographic groups within the Black population were particularly vulnerable.

Other findings include:

- Young Black children (age 0–4 Black Alone or in Combination) had a net undercount rate of 6.3% in the 2010 Census and this is about 50% higher than the overall net undercount rate for young children.
- The net undercount rate for Black males age 18–49 was very high (7.6% based on 2012 DA).
- The net omissions rate for Black male renters 30–49 was extremely high at 19.7%.
- While the net undercount for the Black population has fallen dramatically since 1940, the differential undercount between Blacks and Non-Blacks has only decreased slightly.

References

Choldin, H. (1994). *Looking for the last %. The controversy over census undercounts*. New Brunswick, NJ: Rutgers University Press.

Fay, R. E., Passel, J. S., Robinson, J. G., with Assistance from Cowan, C. D. (1988). *The coverage of the population in the 1980 census*. U.S. Census of Population and Housing, Evaluation, and Research Reports, PHC80-E4. Washington, DC: U.S. Census Bureau.

Fein, D. J. (1989). *The social sources of census omissions: Racial and ethnic differences in omissions rates in recent U.S. censuses*. Princeton, NJ.: Dissertation in Department of Sociology, Princeton University.

Heer, D. M. (Ed). (1968). *Social statistics and the city*. Cambridge, MA: Joint Center for Urban Studies of Massachusetts Institute of Technology and Harvard University.

Hill, R. B. (1975). *Estimating the 1970 census undercount for state and local areas*. Washington, DC: National Urban League.

O'Hare, W. P. (2015). The undercount of young children in the U.S. Decennial census. Springer Publishers.

O'Hare, W. P., Robinson, J. G., West, K., & Mule, T. (2016). Comparing the U.S. Decennial census coverage estimates for children from the demographic analysis and coverage measurement surveys. *Population Research and Policy Review*, 35(5), 685–704.

Passel, J. (1991). *Alternative estimates of Black birth corrected for under-registration: 1935–1980*. Washington, DC: The Urban Institute.

Robinson J. G., Das Gupta, P., & Ahmed, B. (1990, August). A case study in the investigation of errors in estimates of coverage based on demographic analysis: black adults aged 34–54 in 1980. Paper presented at the American Statistical Association Annual Conference. Anaheim, CA.

Robinson J. G. (1997). The differential undercount of adult black men: is it a myth? Internal Census Bureau Memorandum, November 3, Washington DC.: U.S. Census Bureau.

Robinson, J. G. (2010). *Coverage of population in Decennial Census 2000 based on demographic analysis: The history behind the numbers*. Decennial Census Bureau, Working Paper No. 91, available online at http://www.Census.gov/population/www/documentation/twps0091/twps0091.pdf.

U.S. Census Bureau. (1991, June 13). Summary table comparing the census, post enumeration survey, and demographic analysis estimates of the U.S. Resident Population on April 1, 1990. By J. Gregory Robinson Memorandum for Paula J. Schneider, Chief, Population Division.

U.S. Census Bureau. (2012). *2010 components of census coverage for race groups and hispanic origin by age, sex, and tenure in the United States. DSSD 2010 Census Coverage Measurement Memorandum Series #2010-E-51*. Washington, DC: U.S. Census Bureau.

U.S. Office of Management and Budget. (1997, October 30). *Revisions to the standards for the classification of federal data on race and ethnicity*. Statistical Policy Directive 15, Federal Register Notice.

U.S. Office of Management and Budget. (2001) Guidance on aggregation and allocation of data on race for use in civil rights monitoring and enforcement.

Velkoff, V. (2011, March). *Demographic evaluation of the 2010 census*. Paper presented at the 2011 population association of America annual conference. Washington, DC.

Valentine, C. A., & Valentine, B. L. (1971). Missing men: A comparative methodological study of under enumeration and related problems. Unpublished.

West, K., Devine, J., & Robinson, J. G. (2014). *An assessment of historical demographic analysis estimates for the Black Male cohorts of 1935–39*. Boston, MA: Paper presented at the Annual Meeting of the American Statistical Association.

Open Access This chapter is licensed under the terms of the Creative Commons Attribution 4.0 International License (http://creativecommons.org/licenses/by/4.0/), which permits use, sharing, adaptation, distribution and reproduction in any medium or format, as long as you give appropriate credit to the original author(s) and the source, provide a link to the Creative Commons license and indicate if changes were made.

The images or other third party material in this chapter are included in the chapter's Creative Commons license, unless indicated otherwise in a credit line to the material. If material is not included in the chapter's Creative Commons license and your intended use is not permitted by statutory regulation or exceeds the permitted use, you will need to obtain permission directly from the copyright holder.

Chapter 9
Census Coverage of the Asian Population

Abstract In the 2010 Census, the net undercount for Asians Alone or in Combination was so small it rounds to zero compared to a net overcount of 0.8% for the Non-Hispanic White Alone population. Asians also had a relatively low omissions rate in the 2010 Census. The omissions rate for Asians Alone or in Combination (5.3%) was slightly higher than the rate for Non-Hispanic White Alone (3.8%). The Asian age/sex group with the highest net undercount rate was Asian Alone or in Combination males age 18–29 who had a net undercount rate of 2.2% in 2010.

9.1   Introduction

Studying the recent history of U.S. Census coverage of Asians is important because they are one of the fastest growing major race/ethnic groups in the country. Based on Census Bureau population estimates, the Asian Alone or in Combination population grew by 21% between 2010 and 2016 (from 17.3 million in 2010 to 20.9 million in 2016).

In reporting Census data on the Asian population, it is very important to be clear about how the group is defined. Starting in the 2000 Census, people were allowed to select more than one race in the Census questionnaire (U.S. Office of Management and Budget 1997) and race is often shown two different ways in Census Bureau reports. One category is the number of people who only select Asian (referred to as Asian Alone) and second a category is those in the first category plus those who select Asian along with at least one other race (referred to as Asian Alone or in Combination). In the 2016 American Community Survey conducted by the Census Bureau there were 17.6 million people who marked Asian Alone and 20.9 million who marked Asian Alone or Asian and some additional race. Asian Alone or in Combination is the primary population examined in this Chapter. This is the most inclusive definition and its use here is consistent with advice of the U.S. Office of Management and Budget (2001).

The Demographic Analysis method does not produce data on Census coverage for Asians, so all the data in this Chapter come from the Census Bureau's Dual-Systems Estimates (DSE) method.

© The Author(s) 2019
W. P. O'Hare, *Differential Undercounts in the U.S. Census*,
SpringerBriefs in Population Studies, https://doi.org/10.1007/978-3-030-10973-8_9

Table 9.1 Net coverage rates for Asian alone or in combination and Non-Hispanic White Alone populations in the 2010 census by age and sex

	Asian alone or in combination	Non-Hispanic White Alone
Total	0.0	**0.8**
Age 10–17	0.6	**1.8**
Age 18–29 males	−1.2	**1.5**
Age 18–29 females	0.8	**1.1**
Age 30–49 males	**−2.2**	**−2.1**
Age 30–49 females	0.0	**0.7**
Age 50+ males	1.1	**0.6**
Age 50+ females	1.0	**2.2**

Source US Census Bureau (2012). 2010 Components of census coverage for Race groups and Hispanic origin by age, sex, and tenure in the United States. DSSD 2010 Census Coverage Measurement Memorandum Series #2010-E-51, US Census Bureau, Washington, DC. Table C

Note The directionality of undercounts and over counts have been reversed from the original publication in order to keep them consistent within this publication in other words, a negative sign implies an undercount

Figures in **BOLD** are statistically significantly different from zero

9.2   2010 Census Coverage of Asians Alone or in Combination by Age and Sex

Table 9.1 shows net undercount rates from DSE for Asians Alone or in Combination and Non-Hispanic Whites Alone broken down by age and sex. Data for the youngest age group (0–9) are not presented in Table 9.1 because there is strong evidence that the coverage estimates of young children in the DSE in 2010 are problematic (O'Hare et al. 2016).

In the 2010 Census, the net undercount rate for all Asians Alone or in Combination was near zero (Table 9.1). By comparison, there was a statistically significant net overcount of Non-Hispanic White Alone population of 0.8%.

As with most racial and ethnic groups examined here, the net undercount rate for the total Asian Alone or in Combination population masks some important differences by age and sex. However, within the Asian population, the differences by age and sex are not as pronounced as they are for Black or Hispanic populations.

The highest net undercount rate in Table 9.1 is for males age 30–49. The net undercount rate for Asians Alone or in Combination males in this age range is 2.2% and this is the only net undercount estimate for Asian Alone or in Combination that is statistically significantly different than zero. The relatively high net undercount rate of Asian Alone or in Combination males age 30–49 is consistent with most other race and Hispanic groups, where males in this age group have high net undercount rates.

Table 9.2 Omissions rates for Asian Alone or in combination and Non-Hispanic White Alone populations in the 2010 census by age and sex

	Asian Alone or in Combination	Non-Hispanic White Alone
Total	5.3	3.8
Age 10–17	3.3	3.1
Age 18–29 males	8.4	6.6
Age 18–29 females	8.4	6.2
Age 30–49 males	7.8	6.2
Age 30–49 females	4.4	3
Age 50+ males	3.7	3.5
Age 50+ females	3.6	1.7

Source U.S. Census Bureau (2012). DSSD 2010 CENSUS COVERAGE MEASURMENT MEMORANDUM SERIES #2010-E-51, Table C

9.3 Census 2010 Omissions Rates for Asians Alone or in Combination

Recall that the net Census undercount rate is a balance between people omitted and those included erroneously (mostly double counted). The omissions rate captures the share of a group missed in the Census. DSE is the only method that shows omissions rates.

In many ways the omissions rate is a more meaningful statistic because in the net undercount calculation, omissions can be cancelled out by erroneous inclusions or double counting. A net undercount of zero could be the result of no one missed and no one double counted, or, for example, 10% missed, and 10% double counted.

Examination of omissions rates are particularly important for the Asian Alone or in Combination population because the net undercount of zero could leave people with the impression that no Asians were missed in the 2010 census. The data below show that is not the case.

Table 9.2 shows omissions rates for Asians Alone or in Combination and Non-Hispanic White Alone. The omissions rate for Asians Alone or in Combination (5.3%) is higher than the Non-Hispanic White Alone population (3.8%). Omissions rates vary by age and sex but in each age/sex group Asians Alone or in Combination have higher omissions rates than Non-Hispanic Whites Alone. To a large extent the omissions rates by age and sex among Asians Alone or in Combination reflect the same demographic pattern as the net undercount rates.

Like the results for net undercount rates, Asian Alone or in Combination males age 18–49 had relatively high omissions rates. The omissions rate for both Asians Alone or in Combination males and females age 18–29 was 8.4%. The omissions rate for Asian males age 30–49 (7.8%) is relatively high but the rate for Asian Alone or in Combination females in this age range is only 4.4%. The omissions rates for older Asians Alone or in Combination (age 50 and over) are relatively low.

Table 9.3 2010 census net undercount rates and omissions rates for Asian Alone or in combination and Non-Hispanic White Alone by tenure

		Asian Alone or in combination	Non-Hispanic White Alone
Percent undercount	Population living in owner-occupied housing units	0.0	**0.8**
	Population living in renter-occupied housing units	0.0	0.9
Percent omissions	Population living in owner-occupied housing units	4.1	3.0
	Population living in renter-occupied housing units	7.4	6.4

Source US Census Bureau (2012). 2010 components of census coverage for Race Groups and Hispanic origin by age, sex, and tenure in the United States. DSSD 2010 Census Coverage Measurement Memorandum Series #2010-E-51, US Census Bureau, Washington, DC. Table B
A negative sign reflects a net undercount. The signs here are reversed from the source report in order to keep directionality consistent within this publication
Figures in **BOLD** are statistically significantly different from zero

9.4   Differential Undercounts of Asians by Tenure

Table 9.3 shows net undercount rates and omissions rates from the 2010 Census DSE analysis for the population in owner-occupied housing units and for the population in renter-occupied housing units among Asians Alone or in Combination and Non-Hispanic White Alone. The net undercount rates for Asians Alone or in Combination was virtually identical (zero) for both the population living in owner-occupied housing units and those living rental housing units. Interestingly, the net undercount rates for renters and owners are nearly identical for the Non-Hispanic White Alone population as well. This is one of the few places where tenure doesn't seem to make a difference in net undercount rates.

Omissions rates are somewhat higher for Asian Alone or in Combination renters than for homeowners. For both owners and renters, the omissions rates for Asians Alone or in Combination are about one percentage point higher than those of Non-Hispanic White Alone.

Table 9.4 Net undercount rates for Non-Hispanic Asian Alone and Non-Hispanic White Alone; 1990, 2000, 2010

	2010	2000	1990
US Total	0.0	0.5	−1.6
Non-Hispanic White	0.8	1.1	−0.7
Non-Hispanic Asian	−0.1	0.8	−2.4

Source US Census Bureau (2012). 2010 census coverage measurement estimation report: summary of estimates of coverage for persons in the United States. DSSD 2010 Census Coverage Measurement Memorandum Series #2010-G-01, US Census Bureau, Washington, DC. Table 7
[a]A negative sign reflects a net undercount. The signs here are reversed from the source report in order to keep directionality consistent within this publication
Note in this table the race categories are race alone, not race alone or in combination. The data are presented this way to make the 2010 categories consistent with the 1990 categories
Figures in **BOLD** are statistically significantly different than zero

9.5   Trend Data from 1990 to 2010

Table 9.4 shows net undercount rates for 1990, 2000 and 2010 Censuses for Asians Alone and Non-Hispanic Whites Alone based on the Census Bureau's DSE method. Asian Alone is the category used here because that is the only one available in 1990.

Trends in the net undercount rates for both Asians Alone and Non-Hispanic Whites Alone have been inconsistent from 1990 to 2010. For Asians, there was a relatively high net undercount rate in 1990 (2.4%) but in the 2000 Census the net undercount rate for Asians (0.8%) was not statistically significantly different than zero. In 2010, there was a net undercount of 0.1% for Asians Alone compared to an overcount of 0.8 for Non-Asian White Alone population.

The undercount differential between Asians Alone and Non-Hispanic White Alone population fell a little between 1990 and 2000 but increased slightly between 2000 and 2010.

9.6   Census Coverage of Asian Subgroups

In the Census coverage calculations done by the Census Bureau, Asians are treated as a single population. But there are several distinctions that should be made in the Asian population with respect to the probability of being missed in the Census.

There are a great variety of languages, religions, and customs among Asian subgroups (U.S. Census Bureau 2012). Among the Asians counted in the Census Bureau's 2016 American Community Survey there were 4 million Chinese, 3.8 million Asian Indians, 2.8 million Filipinos, 1.8 million Vietnamese, 1.4 million Koreans, and millions of Asians from other countries or ethnic backgrounds. The social, economic, and cultural differences among these subgroups of Asians suggest that they may differ in the levels of Census coverage. O'Hare (2017) shows how undercount-related socioeconomic measures vary for Asian subgroups in California.

Part of the diversity among Asians is linked to the large number of immigrants in this population. The 2016 American Community Survey indicates 66% of the Asian population in the U.S. is foreign-born. In recent years, the number of immigrants from some Asian countries has been higher than any other nation. In 2016, the number of immigrants from India (175, 100) and China/Hong Kong (160, 400) outnumbered those from Mexico (150, 400) (Zong et al. 2018). Consequently, a large and growing segment of the Asian population in the U.S. have many of the characteristics that make recent immigrants difficult to enumerate in the Census (Massey 2014). According to Jensen et al. (2015, p. 1) "The foreign-born, especially recent immigrants, are believed to be a hard-to-count group which increases the likelihood of coverage error for this population." It is also important to note that censuses in some other countries are not always viewed positively and immigrants may bring their negative views about census-taking with them when they move to the U.S.

Given the fact that a substantial portion of the Asian population in the U.S. are recent immigrants from many different Asian counties, they fall into the category of cultural and linguistic minorities. Harkness et al. (2014, p. 245) state, "Cultural and linguistic minorities can be hard to survey, either as the target population or as a subpopulation of a general population survey." The U.S. General Accountability Office (2018, p. 5) also recognizes Cultural and Linguistic minorities as a hard-to-count group in the Census.

A substantial number of Asian immigrants are undocumented immigrants. One estimate indicates there are 1.5 million undocumented immigrants from Asian in the U.S. (Pew Research Center 2017). Evidence suggests the undocumented populations have higher net undercount rates than other immigrants (Warren and Warren 2013; Van Hook et al. 2014).

Given the large number of recent immigrants and undocumented immigrants in the Asian population, the last-minute addition of a question on citizenship to the 2020 Census is likely to have implications for the count of Asians in the 2020 Census (Ross 2018; Barabba and Flynn 2018; Meyers and Goerman 2018, U.S. Census Bureau 2017a, b). See Chap. 15 for more information on this issue.

There are no direct estimates of Census coverage rates for subgroups of Asians produced by the Census Bureau, but the heterogeneity noted above suggests there are likely to be substantial differences in census coverage among Asian subgroups. At least in part, this heterogeneity is why some observers fear Asians may be undercounted in the 2020 Census (Fuchs 2017; The Leadership Conference Education Fund 2107; Kang 1990; Horikoshi and Minnis 2018).

9.7 Summary

While the net undercount for Asian Alone or in Combination is very low in the 2010 Census, there was an omissions rate of 5.3% for Asians Alone or in Combination, some age/sex groups within the Asian population experienced high net undercounts and omissions rates. In addition, it is likely that the Census coverage rates for some Asian subgroups are different than the Census coverage for all Asians.

References

Barabba, V., & Flynn, K. H. (2018, January 30). Ensure Everyone is Counted. *U.S. News.*

Fuchs, C. (2017, November 29). As census approaches, some advocates worried Asian Americans could be undercounted. *NBC Asian American.*

Harkness, J., Stange, M., Cibelli, K. L., Mohler, P., & Pennell, B. (2014). Surveying cultural and linguistic minorities. In R. Tourangeau, B. Edwards, T. P. Johnson, K. M. Wolter, & N. Bates (Eds.), *Hard-to-survey populations* (pp. 245–269). Cambridge, England: Cambridge University Press.

Horikoshi, N., & Minnis, T. A., (2018). Asian American and Pacific Islander boys and men: The risk of being missed in the U.S. 2020 Census. http://www.risebmoc.org/issues/download-post6.

Jensen, E. B., Bhaskar, R., & Scopilliti, M. (2015, June). Demographic analysis 2010: Estimates of coverage of the foreign-born population in the American Community survey. U.S. Census Bureau. Population Division, U.S. Census Bureau Working Paper No. 103.

Kang, T. S. (1990, March). An ethnography of Koreans in Queens, New York and elsewhere in the United States. Ethnographic Exploratory Research Paper No. 8. Washington, DC: Undercount Behavioral Research Group, Bureau of the Census.

Massey, D. (2014). Challenges to surveying immigrants. In R. Tourangeau, B. Edwards, T. P. Johnson, K. M. Wolter, & N. Bates (Eds.), *Hard-to-survey populations* (pp. 270–292). Cambridge, England: Cambridge University Press.

Meyers, M., & Goerman, P. (2018). Respondents confidentiality concerns in multilingual pretesting studies and possible effects on response rates and data quality for the 2020 Census. Paper delivered at the annual Conference of the American Association of Public Opinion Research, Denver CO., May 16–19.

O'Hare, W. P., Robinson, J. G., West, K., & Mule, T. (2016). Comparing the U.S. Decennial census coverage estimates for children from the demographic analysis and coverage measurement surveys. Forthcoming in *Population Research and Policy Review.* https://doi.org/10.1007/s11113-016-9397-x.

O'Hare, W. P. (2017). *Counting all Californians in the 2020 census.* Washington, DC: The Census Project.

Pew Research Center. (2017). Undocumented immigrants, Available at http://www.pewresearch.org/topics/unauthorized-immigration/2017/.

Ross, W. (2018, May 26). Reinstatement of a citizenship question on the 2020 Decennial census questionnaire. Memorandum from Secretary of Commerce Wilber Ross to Undersecretary of Commerce Karen Dunn Kelley.

The Leadership Conference Education Fund. (2017). "Will you count? Asian Americans and Native Hawaiians and Pacific Islanders (NHPIs) in the 2020 Census. Available at http://civilrightsdocs.info/pdf/Census/2020/Fact-Sheet-AA-NHPI-HTC.pdf.

U.S. General Accountability Office. (2018, July). 2020 census: Actions needed to address challenges to enumerating hard-to-count groups. Report to Congressional Requesters, GAO-18-599.

U.S. Office of Management and Budget. (1997). *Revisions to the standards for the classification of federal data on race and ethnicity*. Statistical Policy Directive 15, Federal Register Notice October 30, 1997, Available on line at http://www.Whitehouse.gov/omb/fedreg_1997standards.

U.S. Office of Management and Budget. (2001). Guidance on aggregation and allocation of data on race for use in civil rights monitoring and enforcement.

U.S. Census Bureau (2012). The Asian Population: 2010. 2010 Census Briefs, C2010BR-11, Elizabeth M. Hoeffel, Sonya Rostogi, and Hasan Shahid. Washington, DC: U.S. Census Bureau.

U.S. Census Bureau. (2017a, September 20). Respondent confidentiality concerns. Memorandum for Associate Directorate for Research and Methodology (ADRM) From Center for Survey Measurement (SCM).

U.S. Census Bureau. (2017b, November 2). Respondent confidentiality concerns and possible effects on response rates and data quality for the 2020 census. Presentation by Mikelyn Meyers at the Census Bureau National Advisory Committee Meeting.

Van Hook, J., Bean, F., Bachmeier, J. D., & Tucker, C. (2014). Recent trends in coverage of the Mexican-born population of the United States: Results from applying multiple methods across time. *Demography, 51*(3), 699–726.

Warren, R., & Warren, J. R. (2013). Unauthorized immigration to the United States: Annual estimates and components of changes, by State, 1990 to 2010. *International Migration Review, 47*(2), 296–329.

Zong, J., Batalova, J., & Hallock, J. (2018). *Frequently requested statistics on immigrants and immigration in the United States*. Washington, DC: Migration Policy Institute.

Open Access This chapter is licensed under the terms of the Creative Commons Attribution 4.0 International License (http://creativecommons.org/licenses/by/4.0/), which permits use, sharing, adaptation, distribution and reproduction in any medium or format, as long as you give appropriate credit to the original author(s) and the source, provide a link to the Creative Commons license and indicate if changes were made.

The images or other third party material in this chapter are included in the chapter's Creative Commons license, unless indicated otherwise in a credit line to the material. If material is not included in the chapter's Creative Commons license and your intended use is not permitted by statutory regulation or exceeds the permitted use, you will need to obtain permission directly from the copyright holder.

Chapter 10
Census Coverage of American Indians and Alaskan Natives

Abstract Overall, the net undercount of American Indians and Alaskan Natives Alone or in Combination is relatively low (0.2%) but some subgroups within this population have high net undercount and omissions rates. The omissions rate for American Indians and Alaskan Natives Alone or in Combination was 7.6% which is double the rate for Non-Hispanic White Alone (3.8%). Also, Census coverage for American Indians and Alaskan Natives living on reservations was much worse than for this population living elsewhere. The net undercount rate for American Indians Alone and Alaskan Natives or in Combination living on reservations was 4.9% in the 2010 Census.

10.1   Introduction

In reporting Census data on the American Indian and Alaskan Native population it is very important to be clear about how the group is defined. Starting in the 2000 Census, people were allowed to select more than one race in the Census questionnaire (U.S. Office of Management and Budget 1997). Given this option, race is often shown two different ways in Census Bureau reports. One category is the number of people who only select American Indian and Alaskan Native (referred to as American Indians and Alaskan Natives Alone) and a second category includes everyone in the first category as well as those who select American Indian and Alaskan Native along with at least one other race (referred to as American Indians and Alaskan Native Alone or in Combination).

The Census Bureau population estimates for 2016 show 4.1 million people identified as American Indian and Alaskan Native Alone and 6.7 million selected American Indian and Alaskan Native Alone or in Combination. American Indian and Alaskan Natives Alone or in Combination is the primary definition for the population examined in this Chapter. This is the most inclusive definition and using it in this Chapter is consistent with advice of the U.S. Office of Management and Budget (2001).

The Demographic Analysis (DA) method does not produce data on Census coverage for American Indians and Alaskan Natives, because this category has not been used consistently in birth and death certificates over time or across states. All the

© The Author(s) 2019   101
W. P. O'Hare, *Differential Undercounts in the U.S. Census*,
SpringerBriefs in Population Studies, https://doi.org/10.1007/978-3-030-10973-8_10

data in this section come from the Census Bureau's Dual-Systems Estimates (DSE) method. Keep in mind the sample size of American Indians and Alaskan Natives in the Post-Enumeration Survey used in the DSE method is smaller than most other groups so small differences between rates may not be significant.

10.2   Undercount of American Indians and Alaskan Natives

Table 10.1 shows net undercount rates from DSE for American Indians and Alaskan Natives Alone or in Combination and Non-Hispanic Whites Alone broken down by age and sex. Data for the children under age 10 are not provided in Table 10.1 because there is strong evidence that the DSE coverage estimates of young children are problematic (O'Hare et al. 2016).

In the 2010 Census, there was a small net undercount (0.2%) for American Indians and Alaskan Natives Alone or in Combination which was not statistically significantly different from zero. By comparison, there was a statistically significant net overcount of Non-Hispanic White Alone population of 0.8%.

As with all the racial and ethnic groups examined in this publication, the net undercount rate for the total population masks some important differences among age/sex subgroups. However, within the American Indian and Alaskan Native Alone or in Combination population, the differences by age and sex are not as pronounced as they are in the Black and Hispanic populations.

Table 10.1 Net undercount rates for American Indians and Alaskan Natives Alone or in Combination population and Non-Hispanic White Alone population in the 2010 census by age and sex

	American Indians and Alaskan Natives Alone or in Combination	Non-Hispanic White Alone
Total	−0.2	**0.8**
Age 10–17	1.1	**1.8**
Age 18–29 male	−1.9	**1.5**
Age 18–29 female	**−2.6**	**1.1**
Age 30–49 male	**−3.1**	**−2.1**
Age 30–49 female	0.4	**0.7**
Age 50+ male	1.5	**0.6**
Age 50+ female	**4.6**	**2.2**

Source U.S. Census Bureau (2012a). "2010 Components of Census Coverage for Race Groups and Hispanic Origin by Age, Sex, and Tenure in the United States". DSSD 2010 CENSUS COVERAGE MEASUREMENT MEMORANDUM SERIES#2010-E-51, U.S. Census Bureau, Washington, DC
A negative sign reflects a net undercount. The signs here are reversed from the source report in order to keep directionality consistent within this publication
Figures in **BOLD** are statistically significantly different from zero

The highest net undercount rate in Table 10.1 is for American Indian and Alaskan Native Alone or in Combination males age 30–49 where the net undercount rate is 3.1% which is statistically significant. There is also a high statistically significant net undercount rate for females age 18–29 (2.6%). On the other hand, there is a large net overcount of American Indian and Alaskan Native Alone or in Combination females over age 50 (4.6%). It is not clear why older American Indians and Alaskan Natives Alone or in Combination females have such a high net overcount rate. These are the only net coverage estimates for American Indians and Alaskan Natives Alone or in Combination that are statistically significantly different than zero.

10.3 Census Coverage on Reservations

One important distinction among American Indians and Alaskan Natives Alone or in Combination with respect to the accuracy of Census counts is whether they live on an Indian Reservation. Data indicate American Indians and Alaskan Natives Alone or in Combination living on Indian reservations have much higher net undercounts than those living elsewhere in the country (U.S. Census Bureau 2012b, Table 7). In the 2010 Census, there were 571,000 American Indians or Alaskan Natives living on Indian Reservations (U.S. Census Bureau 2012b, Table 9).

The net undercount in the 2010 Census for American Indians and Alaskan Natives Alone or in Combination living on reservations was 4.9%. For American Indians Alone and Alaskan Natives or in Combination living in elsewhere there was a net overcount of 2.0%. The omissions rate for American Indians and Alaskan Natives Alone or in Combination living on reservations was 13.7% (U.S. Census 2012b, Table 9).

The high net undercount and omissions rates of American Indians and Alaskan Natives living on reservations may reflect the presence of some of the same hard-to-count (HTC) characteristics (high poverty and unemployment rates, unconventional housing, and low levels of literacy, for example) that exist in many HTC urban neighborhoods.

10.4 Omissions Rates for American Indians and Alaskan Natives

Recall that the net Census undercount rate is a balance between people omitted and those included erroneously (mostly double counted). The omissions rate captures the share of a group missed in the Census. DSE is the only method that shows omissions rates.

Table 10.2 Omissions rates for American Indians and Alaskan Natives Alone or in Combination and Non-Hispanic White Alone populations in the 2010 census by age and sex

	American Indians and Alaskan Natives Alone or in Combination	Non-Hispanic White Alone
Total	7.6	3.8
Age 10–17	8.1	3.1
Age 18–29 males	11.2	6.6
Age 18–29 females	10.3	6.2
Age 30–49 males	9.9	6.2
Age 30–49 females	5.5	3
Age 50+ males	5.9	3.5
Age 50+ females	3.2	1.7

Source U.S. Census Bureau (2012a). "2010 Components of Census Coverage for Race Groups and Hispanic Origin by Age, Sex, and Tenure in the United States". DSSD 2010 CENSUS COVERAGE MEASUREMENT MEMORANDUM SERIES#2010-E-51, U.S. Census Bureau, Washington, DC. Table C

In many ways the omissions rate is a more meaningful statistic than the net undercount rate because in the net undercount calculation omissions can be cancelled out by erroneous inclusions or double counting. A net undercount of zero could be the result of no one missed and no one double counted, or for example, ten percent missed, and ten percent double counted.

Examination of omissions rates are particularly important for the American Indians and Alaskan Natives Alone or in Combination population because the very low net undercount could leave people with the impression that almost no American Indians and Alaskan Natives Alone or in Combination were missed in the 2010 Census. The data below show that is not the case.

Omissions rates show much bigger differences between American Indian and Native Alaskans Alone or in Combination and Non-Hispanic Whites Alone than were seen in comparisons of net undercount rates. Table 10.2 shows omissions rates by age and sex for American Indians and Alaskan Natives Alone or in Combination and the Non-Hispanic Whites Alone population. The omissions rate for American Indians and Alaskan Natives Alone or in Combination (7.6%) is double that of the Non-Hispanic White Alone population (3.8%).

Data in Table 10.2 show that the omissions rates for American Indians and Alaskan Natives Alone or in Combination are higher than the corresponding cell for Non-Hispanic Whites Alone in every case. The omissions rates for American Indians and Alaskan Natives Alone or in Combination varies by age and sex and to a large extent the omissions rates reflect the same pattern as the net undercount rates.

Like the results for net undercount rates, American Indian and Alaskan Native Alone or in Combination males age 18–49 had higher omissions rates than any other sex/age subgroup. For young adults (age 18–29) the omissions rates were high for both males and females. The omissions rate for American Indians and Alaskan

Table 10.3 2010 census net undercount rates and omissions rates for American Indians and Alaskan Natives Alone or in Combination and Non-Hispanic White Alone by Tenure

		American Indians and Alaskan Natives Alone or in Combination	Non-Hispanic White Alone
Percent undercount	Population living owner-occupied housing units	1.3	**0.8**
	Population living renter-occupied housing units	**−1.9**	0.9

Source U.S. Census Bureau (2012a). "2010 Components of Census Coverage for Race Groups and Hispanic Origin by Age, Sex, and Tenure in the United States". DSSD 2010 CENSUS COVERAGE MEASUREMENT MEMORANDUM SERIES#2010-E-51, U.S. Census Bureau, Washington, DC. Table B
A negative sign reflects a net undercount. The signs here are reversed from the source report in order to keep directionality consistent within this publication
Figures in **BOLD** are statistically significantly different from zero

Natives Alone or in Combination males age 18–29 was 11.2% and for females age 18–29 it was 10.3%. The omissions rate for males age 30–49 was 9.9%. Older American Indians and Alaskan Natives Alone or in Combination (age 50 and over) omissions rates were lower.

10.5 Coverage of American Indians and Alaskan Natives by Tenure

Table 10.3 shows net coverage rates and omissions rates from the 2010 Census DSE analysis for the population living in owner-occupied housing units and the population living in renter-occupied housing units among American Indians and Alaskan Natives Alone or in Combination compared to the Non-Hispanic White Alone population.

In terms of net undercount rates, the population of American Indians and Alaskan Natives Alone or in Combination living in owner-occupied housing units had a net overcount of 1.3% which is slightly higher than that of Non-Hispanic Whites Alone (0.8%). On the other hand, American Indian and Alaskan Natives Alone or in Combination living in rental-occupied housing units had a 1.9% net undercount rate compared to a 0.9% net overcount for Non-Hispanic Whites Alone living in rental housing units.

Table 10.4 Net undercount rates for American Indians Alone, Non-Hispanic White Alone in the U.S. Census, 1990, 2000, and 2010

	2010	2000	1990
	Undercount rate	Undercount rate	Undercount rate
American Indians Alone living on reservations	**−4.9**	0.9	**−12.2**
American Indians Alone living not living on reservations	2.0	−0.6	**−0.7**
Non-Hispanic White	**0.8**	**1.1**	**−0.7**

Source U.S. Census Bureau (2012a). "2010 Census Coverage Measurement Estimation Report: Summary of Estimates of Coverage for Persons in the United States". DSSD 2010 CENSUS COVERAGE MEASUREMNET MEMORANDUM SERIES#2010-G-01, U.S. Census Bureau, Washington, DC. Table 7

A negative sign reflects a net undercount. The signs here are reversed from the source report in order to keep directionality consistent within this publication

Figures in **Bold** are statistically significantly different than zero

In terms of omissions rates, American Indian and Alaskan Natives Alone or in Combination had higher rates than Non-Hispanic Whites Alone for both owners and renters, but the difference was a little larger for owner.

10.6   Trend Data from 1990 to 2010

Table 10.4 shows net undercount rates for 1990, 2000, and 2010 Censuses for American Indians Alone and Non-Hispanic Whites Alone based on the Census Bureau's DSE method. The race alone category is used in Table 10.4 because that is the only category available in 1990. The data for American Indians Alone has been broken out by whether or not a person was living on an Indian Reservation because there are big differences between people living on a reservation or somewhere else.

The net undercount rates for American Indians Alone on reservations have been wildly inconsistent from 1990 to 2010. For American Indians Alone living on reservations there was a very high net undercount rate in 1990 (12.2%) but in the 2000 Census the there was a small (0.9%) net overcount which is not statistically significantly different from zero. In the 2010 Census there was a statistically significant net undercount rate of 4.9%. It is difficult to discern any trends over time from this data and I am not aware of any explanation for the big swings in Census coverage for American Indians from 1990 to 2010. For American Indians Alone living somewhere other than a reservation there was a small net undercount in 1990 and 2000 but a net overcount in 2010.

10.7 Potential Addition of a Question on Citizenship

It may seem odd to being talking about American Indians and a question on citizenship, but I think it may be more relevant than it first appears. In March 2018, Commerce Secretary Ross (2018) announced that the 2020 Census questionnaire would have a question on citizenship. At the Spring 2018 Meeting of the Census Bureau's National Advisory Committee it was reported that some American Indians would not identify as being citizens of the United States, because they feel they are citizens of their tribal nation. This is consistent with the 2016 American Community Survey, which found 3.3% of American Indian and Alaskan Natives Alone or in Combination reported they were not a U.S. citizen and 6% did not provide a response to this question. The nonresponse rate for the citizenship question is much higher than the nonresponse rates for basic demographic characteristics such as age, sex, and race, which suggests in may be a more sensitive question (O'Hare 2018). This is likely to add another (unforeseen) twist to the last-minute addition of the citizenship question to the 2020 Census questionnaire (Ross 2018; Barraba and Flynn 2018).

10.8 Summary

Overall, the net undercount of American Indians and Alaskan Natives is relatively low (0.2%) but some subgroups within this population have high net undercount and omissions rates. Like most other groups examined in this book, males age 18–49 had the highest net undercount and omissions rates.

In the 2010 Census, the net undercount rate of American Indians living on reservations was very high (4.9%). Many American Indian reservations have some of the same hard-to-count characteristic (high poverty and unemployment rates, unconventional housing, and low levels of literacy, for example) that exist in many hard-to-count urban neighborhoods.

References

Barabba, V. & Flynn, K. H. (January 30, 2018). Ensure everyone is counted. *U.S. News*.

O'Hare, W. P., Robinson, J. G., West, K., & Mule, T. (2016). Comparing the U.S. decennial census coverage estimates for children from the demographic analysis and coverage measurement surveys. *Population Research and Policy Review, 35*(5), 685–703.

O'Hare, W. P. (2018). *Demographic profile of individuals who don't respond to the u.s. census bureau's american community survey question on citizenship*. Washington, DC: Georgetown University Center on Poverty and Inequality.

Ross, W. (2018). *Reinstatement of a Citizenship Question on the 2020 Decennial Census Questionnaire*. Memorandum from Secretary of Commerce Wilber Ross to Undersecretary of Commerce Karen Dunn Kelley, May 26, 2018.

U.S. Census Bureau. (2012a). The American Indian and Alaskan Native population: 2010. In T. Norris, L. P. Vines, & E. M. Hoeffel (Eds.), *2010 census briefs*, *C2010BR-10*, Washington, DC: U.S. Census Bureau.

U.S. Bureau of the Census. (2012b). *2010 Census coverage measurement estimation report: Summary of estimates of coverage for persons in the United States*, DSSD 2010 Census Coverage Measurement Memorandum Series 2012, 2010- G-01. Washington, DC: U.S. Census Bureau.

U.S. Office of Management and Budget. (1997). *Revisions to the Standards for the Classification of Federal Data on Race and Ethnicity*, Statistical Policy Directive 15, Federal Register Notice, October 30.

U.S. Office of Management and Budget. (2001). *Guidance on aggregation and allocation of data on race for use in civil rights monitoring and enforcement*.

Open Access This chapter is licensed under the terms of the Creative Commons Attribution 4.0 International License (http://creativecommons.org/licenses/by/4.0/), which permits use, sharing, adaptation, distribution and reproduction in any medium or format, as long as you give appropriate credit to the original author(s) and the source, provide a link to the Creative Commons license and indicate if changes were made.

The images or other third party material in this chapter are included in the chapter's Creative Commons license, unless indicated otherwise in a credit line to the material. If material is not included in the chapter's Creative Commons license and your intended use is not permitted by statutory regulation or exceeds the permitted use, you will need to obtain permission directly from the copyright holder.

Chapter 11
Census Coverage of the Native Hawaiian or Pacific Islander Population

Abstract In general, the count of Native Hawaiian or Pacific Islanders Alone or in Combination in the U.S. Census is relatively accurate. The net undercount rate for Native Hawaiian or Pacific Islanders Alone or in Combination in 2010 was 1.0% compared to a net overcount of 0.8% for the Non-Hispanic White Alone population. The omissions rate for Native Hawaiian or Pacific Islanders Alone or in Combination (7.9%) is about double the rate for Non-Hispanic White Alone (3.8%). The Native Hawaiian or Pacific Islanders Alone or in Combination demographic group with the highest net undercount rate was males age 18–29 who had a net undercount rate of 8.0% in 2010. This is the only net undercount rate among Native Hawaiian or Pacific Islanders that was statistically significantly different than zero.

11.1 Introduction

In reporting Census data on the Native Hawaiian or Pacific Island population it is very important to be clear about how the group is defined. Starting in the 2000 Census, people were allowed to select more than one race in the Census questionnaire (U.S. Office of Management and Budget 1997) and race is often reported two different ways in Census Bureau reports. One category is the number of people who only select Native Hawaiian or Pacific Islander (referred to as Native Hawaiian or Pacific Islander Alone) and the second category is those in the first group plus those who select Native Hawaiian or Pacific Islander along with at least one other race (referred to as Native Hawaiian or Pacific Islander Alone or in Combination).

There is a substantial difference between the number of people who are in the Native Hawaiian or Pacific Islander Alone category and those in the Native Hawaiian or Pacific Islander Alone or in Combination category. In the 2016 American Community Survey conducted by the Census Bureau there were about 600,000 people who marked Native Hawaiian or Pacific Islander Alone and 1.4 million who marked Native Hawaiian or Pacific Islander and some additional race. Except for data on trends over time, Native Hawaiians or Pacific Islanders Alone or in Combination is the population examined in this Chapter. This is the most inclusive definition and consistent with advice of the U.S. Office of Management and Budget (2001).

© The Author(s) 2019

W. P. O'Hare, *Differential Undercounts in the U.S. Census*,

SpringerBriefs in Population Studies, https://doi.org/10.1007/978-3-030-10973-8_11

It is also noteworthy that the number of Native Hawaiians or Pacific Islanders Alone or in Combination increased by 40% between 2000 and 2010 (U.S. Census Bureau 2012). About half of the Native Hawaiian or Pacific Islander Alone or in Combination population resides in Hawaii or California.

This population is not only increasing rapidly but given the unique history and evolution of Native Hawaiians or Pacific Islanders in the U.S. population, they fall into the category of cultural and linguistic minorities discussed by Harkness et al. (2014). Cultural and linguistic minority are often considered a hard-to-count population which means they are more difficult to enumerate accurately in the census.

The Demographic Analysis method does not produce data on Census coverage for Native Hawaiian or Pacific Islanders, so all the data in this section comes from the Census Bureau's Dual-Systems Estimates (DSE) method. Keep in mind the sample size of Native Hawaiians and Alaskan Natives in the Post-Enumeration Survey used in the DSE method is smaller than most other groups so small differences between rates may not be significant.

11.2   Census Coverage of Native Hawaiian or Pacific Islanders Alone or in Combination

Table 11.1 shows net undercount rates from DSE for Native Hawaiian or Pacific Islanders Alone or in Combination broken down by age and sex. Data for the youngest age groups (0–9) are not presented in Table 11.1 because there is strong evidence that the coverage estimates of young children in the DSE in 2010 are problematic (O'Hare et al. 2016).

In the 2010 Census, the net undercount rate for all Native Hawaiian or Pacific Islanders Alone or in Combination was 1.0% (Table 11.1) and the rate is not statistically significant different from zero. By comparison, there was a statistically significant net overcount of Non-Hispanic White Alone population of 0.8%.

As with the other racial and Hispanic Origin groups examined in this book, the net undercount rate for the total Native Hawaiian or Pacific Islander Alone or in Combination population masks some important differences by age and sex. However, within the Native Hawaiian or Pacific Islander Alone or in Combination population the differences by age and sex are not as pronounced as they are in Black or Hispanic populations.

The highest net undercount in Table 11.1 is for males age 18–29. The net undercount rate for Native Hawaiian or Pacific Islander males in this age range is 8.0% and this is the only net undercount estimate for Native Hawaiians or Pacific Islanders Alone or in Combination that is statistically significantly different than zero. The relatively high net undercount of males age 18–29 is consistent with high net undercounts for this age/sex group in most other race and Hispanic groups. Young adult males are a demographic group that has been identified as difficult to enumerate regardless of race or ethnicity in many countries (Simpson and Middleton 1997).

Table 11.1 Net undercount rates for Native Hawaiians or Pacific Islanders Alone or in Combination and Non-Hispanic White Alone populations in the 2010 census by age and sex

	Native Hawaiian or Pacific Islanders Alone or in Combination	Non-Hispanic White Alone
Total	−1.0	**0.8**
Age 10–17	1.1	**1.8**
Age 18–29 males	**−8.0**	**1.5**
Age 18–29 females	1.9	**1.1**
Age 30–49 males	2.0	**−2.1**
Age 30–49 females	−1.2	**0.7**
Age 50+ males	−0.4	**0.6**
Age 50+ females	−1.4	**2.2**

Source U.S. Census Bureau (2012). "2010 Components of Census Coverage for Race Groups and Hispanic Origin by Age, Sex, and Tenure in the United States". DSSD 2010 CENSUS COVERAGE MEASUREMENT MEMORANDUM SERIES#2010-E-51, U.S. Census Bureau, Washington, DC. Table C
A negative sign reflects a net undercount. The signs here are reversed from the source report in order to keep directionality consistent within this publication
Figures in **BOLD** are statistically significantly different from zero

11.3 Census 2010 Omissions Rates for Native Hawaiian or Pacific Islanders Alone or in Combination and Non-Hispanic Whites Alone

Recall that the net census undercount rate is a balance between people omitted and those included erroneously (mostly double counted). The omissions rate captures the share of a group missed in the Census. DSE is the only method that shows omissions rates.

In many ways the omissions rate is a more meaningful statistic because in the net undercount calculation, omissions can be cancelled out by erroneous inclusions or double counting. A net undercount of zero could be the result of no one missed and not one double counted, or for example, 10% missed, and 10% double counted.

The omissions rate for Native Hawaiian or Pacific Islanders Alone or in Combination (7.9%) is slightly more than double that of the Non-Hispanic White Alone population (3.8%) but omissions rates vary by age and sex. To a large extent the omissions rates among Native Hawaiian or Pacific Islanders Alone or in Combination reflect the same age/sex demographic pattern as the net undercount rates (Table 11.2).

Like the results for net undercount rates, Native Hawaiian or Pacific Islander Alone or in Combination males age 18–29 had a very high omissions rate. The omissions rate for Native Hawaiian or Pacific Islander Alone or in Combination males 18–29 was 15.7% which is almost twice as high as the overall omissions rate for the Native Hawaiian or Pacific Islanders Alone or in Combination population. The omissions

Table 11.2 Omissions rates for Native Hawaiians and Pacific Islanders Alone or in Combination and Non-Hispanic White Alone populations in the 2010 census by age and sex

	Native Hawaiian or Pacific Islander Alone or in Combination	Non-Hispanic White Alone
Total	7.9	3.8
Age 10–17	5.0	3.1
Age 18–29 male	15.7	6.6
Age 18–29 female	7.7	6.2
Age 30–49 male	6.4	6.2
Age 30–49 female	7.2	3.0
Age 50+ male	4.5	3.5
Age 50+ female	7.6	1.7

Source U.S. Census Bureau (2012). "2010 Components of Census Coverage for Race Groups and Hispanic Origin by Age, Sex, and Tenure in the United States". DSSD 2010 CENSUS COVERAGE MEASUREMENT MEMORANDUM SERIES#2010-E-51, U.S. Census Bureau, Washington, DC. Table C

rate for Native Hawaiian or Pacific Islanders Alone or in Combination males age 30–49 (6.4%) is relatively low compared to the overall omissions rate (7.9%).

11.4 Coverage of Native Hawaiian or Pacific Islanders Alone or in Combination by Tenure

Table 11.3 shows net undercount rates and omissions rates from the 2010 Census DSE analysis for the population living in owner-occupied housing units and renter-occupied housing units among Native Hawaiian or Pacific Islanders Alone or in Combination and Non-Hispanic White Alone.

Like most other demographic groups examined in this book, there is a big difference in census coverage between the population living in owner-occupied housing units and those living in rental units. The net undercount rate for the population of Native Hawaiians or Pacific Islanders Alone or in Combination living in renter-occupied housing units is 3.7% (which is statistically significantly different than zero) compared to a net overcount of 1.8% for Native Hawaiians or Pacific Islanders Alone or in Combination living in owner-occupied housing units. The net coverage gap between Native Hawaiian and Pacific Islanders Alone or in Combination and Non-Hispanic Whites is more than 4 percentage points for renters compared to only 1 percentage point for owners.

Native Hawaiian or Pacific Islanders alone or in Combination have higher omissions rates than Non-Hispanic White Alone in both owner and rental situations. Among the population living in owner-occupied housing units, the omissions rate was 5.6% for Native Hawaiian or Pacific Islanders Alone or in Combination and

Table 11.3 2010 census net undercount rates and omissions rates for Native Hawaiians and Pacific Islanders Alone or in Combination and Non-Hispanic White Alone by Tenure

		Native Hawaiians and Pacific Islanders Alone or in Combination	Non-Hispanic White Alone
Percent Undercount (%)	Owners	1.8	**0.8**
	Renters	**−3.7**	0.9
Percent Omissions (%)	Owners	5.6	3.0
	Renters	10.1	6.4

Source U.S. Census Bureau (2012), DSSD 2010 CENSUS COVERAGE MEASUREMENT MEMORANDUM SERIES 2012, 2010-E-51, Table B

Note The directionality of undercounts and overcounts have been reversed from the original table in order to keep them consistent within this publication In order words, a negative sign implies an undercount

Figures in **BOLD** are statistically significantly different from zero

3.0% for Non-Hispanic White Alone. Among renters the omissions rate for Native Hawaiian or Pacific Islanders Alone or in Combination Native Hawaiian or Pacific Islanders Alone was 10.1% compared to 6.4% for Non-Hispanic White Alone. The omissions rate for Native Hawaiians or Pacific Islanders Alone or in Combination for renter (10.1%) is higher than that of owners (5.6%) which again show the impact of home-ownership on Census accuracy.

11.5 Trend Data from 1990 to 2010

Table 11.4 shows net undercount rates for 1990, 2000, and 2010 Censuses for Native Hawaiian or Pacific Islanders Alone and Non-Hispanic Whites Alone based on the Census Bureau's DSE method. Note that the Native Hawaiian or Pacific Islander "Alone" race definition is used in this table to make categories consistent over time.

The net undercount rates for Native Hawaiians or Pacific Islander Alone have decreased steadily over time from 2.4% in 1990, to 2.1% in 2000, to 1.3% in 2010. While the improvement since 1990 is positive it was very small, and it should be noted that Native Hawaiians and Pacific Islanders still have a net undercount while Non-Hispanic Whites have an overcount.

The difference between the census coverage rates of Native Hawaiian or Pacific Islanders has decreased slightly from 3.1 percentage points in 1990 to 2.1 percentage points in 2010. A small positive improvement.

Table 11.4 Net undercount rates for Non-Hispanic Asian and Non-Hispanic White 1990, 2000, and 2010

	2010	2000	1990
Non-Hispanic White	**0.8**	**1.1**	**−0.7**
Native Hawaiian and Pacific Islanders	−1.3	−2.1	**−2.4**

Source U.S. Census Bureau (2012). "2010 Census Coverage Measurement Estimation Report: Summary of Estimates of Coverage for Persons in the United States". DSSD 2010 CENSUS COVERAGE MEASUREMNET MEMORANDUM SERIES#2010-G-01, U.S. Census Bureau, Washington, DC. Table 7

A negative sign reflects a net undercount. The signs here are reversed from the source report in order to keep directionality consistent within this publication

Note In this table the race categories are race alone, not race alone or in combination. The data are presented this way to make the 2010 categories consistent with the 1990 categories

Figures in **BOLD** are statistically significantly different than zero

11.6 Summary

Overall, the Native Hawaiian or Pacific Islander Alone or in Combination population had a small net undercount (1.0%) compared to a small net overcount for the Non-Hispanic White Alone population. Native Hawaiian or Pacific Islander Alone or in Combination males age 18–29 had the highest net undercount rate of any age/sex group in this population and they had the highest omissions rate of any age/sex group. The net undercount rates for Native Hawaiian or Pacific Islanders have decreased from 1990 to 2010 but it should be noted that Native Hawaiians and Pacific Islanders still have a net undercount while Non-Hispanic Whites have an overcount.

References

Harkness, J., Stange, M., Cibelli, K. L., Mohler, P., & Pennell, B. (2014). Surveying cultural and linguistic minorities. In R. Tourangeau, B. Edwards, T. P. Johnson, K. M. Wolter, & N. Bates (Eds.), *Hard-to-survey populations* (pp. 245–269). Cambridge, England: Cambridge University Press.

O'Hare, W. P., Robinson, J. G., West, K., & Mule, T. (2016). Comparing the U.S. decennial census coverage estimates for children from the demographic analysis and coverage measurement surveys. *Population Research and Policy Review*. https://doi.org/10.1007/s11113-016-9397-x.

Simpson, L. & Middleton, E. (1997). *Who is missed by a national census? A review of empirical results from Australia, Britain, Canada, and the USA*, The Cathie Marsh Centre for Census and Survey Research. UK: University of Manchester.

U.S. Census Bureau. (2012). The native Hawaiian or Pacific Islanders population: 2010. In E. M. Hoeffel, S. Rostogi, & H. Shahid (Eds.), *2010 census brief*. Washington, DC: U.S. Census Bureau.

U.S. Office of Management and Budget. (1997). *Revisions to the standards for the classification of federal data on race and ethnicity.* Statistical Policy Directive 15. *Federal Register.* Notice October 30, 1997. Available online at http://www.Whitehouse.gov/omb/fedreg_1997standards.

U.S. Office of Management and Budget. (2001). *Guidance on aggregation and allocation of data on race for use in civil rights monitoring and enforcement.*

Open Access This chapter is licensed under the terms of the Creative Commons Attribution 4.0 International License (http://creativecommons.org/licenses/by/4.0/), which permits use, sharing, adaptation, distribution and reproduction in any medium or format, as long as you give appropriate credit to the original author(s) and the source, provide a link to the Creative Commons license and indicate if changes were made.

The images or other third party material in this chapter are included in the chapter's Creative Commons license, unless indicated otherwise in a credit line to the material. If material is not included in the chapter's Creative Commons license and your intended use is not permitted by statutory regulation or exceeds the permitted use, you will need to obtain permission directly from the copyright holder.

Chapter 12
Undercount Differentials by Tenure

Abstract Over the past several decades the population living in rental housing units has consistently had net undercounts while the population living in owner-occupied housing units had net overcounts. In this Chapter the coverage differentials by tenure are analyzed and the impact of this differential on different sociodemographic groups is explored.

12.1 Introduction

Home ownership is associated with a set of characteristics that often impact the likelihood of being counted in the Census. A lot of the relationships between tenure and other demographic characteristics have already been covered in earlier Chapters of this book but some are repeated here.

The number and share of the population living in rental housing units has increased since 2010. In the 2010 Census, there were 99.5 million people (33% of the total population) living in rental housing units but the 2016 American Community Survey (U.S. Census Bureau 2018a) shows 111 million people living in rental units which represents 35.3% of the population. As noted above, the population living in rental housing units had higher net undercount and higher omissions rates than the population living in owner-occupied housing units. All other things being equal, this trend is one more factor that will make the 2020 Census more difficult that the 2010 Census (see Chap. 15 for more on this topic).

12.2 Census Coverage by Tenure

Table 12.1 shows net undercount rates and omissions rates for the population living in owner-occupied housing units and the population living in renter-occupied housing units in the 2010 Census. Data in Table 12.1 show that people living in owner-occupied housing units had a net overcount (0.6%) and people living in rental units

© The Author(s) 2019

W. P. O'Hare, *Differential Undercounts in the U.S. Census*,
SpringerBriefs in Population Studies, https://doi.org/10.1007/978-3-030-10973-8_12

Table 12.1   2010 census net undercount rates and omissions rates by tenure

Percent undercount	Population living in owner-occupied housing units	**0.6**
	Population living in renter-occupied housing units	**−1.1**
Percent omissions (%)	Population living in owner-occupied housing units	3.7
	Population living in renter-occupied housing units	8.5

Source U.S. Bureau of the Census, 2012, DSSD 2010 CENSUS COVERAGE MEASUREMENT
MEMORANDUM SERIES 2012, 2010-E-51, Table B
A negative sign reflects a net undercount. The signs here are reversed from the source report in
order to keep directionality consistent within this publication

had a net undercount (1.1%). Both figures are statistically significantly different than
zero. The omissions rate for people living in rental units is 8.5% compared to 3.7%
for those living in owner-occupied units.

12.3   Differential Census Coverage by Tenure, Race, and Hispanic Origin

Table 12.2 provides data for net undercount rates by Race, Hispanic Origin, and
Tenure. In general, the population living in rental units has higher net undercount
rates than those living in owner-occupied units. Only one race group of owners
(Black Alone or in Combination) had a net undercount. Among renters, four of the
six Race/Hispanic groups had a net undercount.

For Blacks Alone or in Combination, Native Hawaiian or Pacific Islanders Alone
or in Combination, and for Hispanics, renters have a net undercount of 3% or more,
signaling that these groups are particularly vulnerable to being missed in the Census.

It is worth noting that for Asians Alone or in Combination the net coverage rate
for both renters and owners is zero and the net overcount rates for Non-Hispanic
Whites Alone owners and renters are nearly identical (0.8% for owners compared to
0.9% for renters). The average income for Asians and Non-Hispanic Whites is above
average, which suggests renters who have the means to secure better rental property
and may not be disadvantaged in the Census.

12.4   Differential Omissions Rates by Tenure, Race, and Hispanic Origin

Recall that the net Census undercount rate is a balance between people omitted and
those included erroneously (mostly double counted). The omissions rate captures the
share of a group missed in the Census. DSE is the only method that shows omissions
rates.

Table 12.2 2010 census net undercount rates by race, Hispanic Origin, and tenure

	Percent undercount	
	Population living in owner-occupied housing units	Population living in rental housing units
Total	**0.6**	**−1.1**
Non-Hispanic White Alone	**0.8**	0.9
Black Alone or in Combination	**−0.9**	**−3.0**
American Indian and Alaskan Natives Alone or in Combination	1.3	**−1.9**
Asian Alone or in Combination	0.0	0.0
Native Hawaiian or Pacific Islander Alone or in Combination	1.8	−3.7
Hispanic Origin	0.3	**−3.3**

Source U.S. Bureau of the Census, 2012, DSSD 2010 CENSUS COVERAGE MEASUREMENT MEMORANDUM SERIES 2012, 2010-E-51, Tables A and B
A negative sign reflects a net undercount. The signs here are reversed from the source report in order to keep directionality consistent within this publication
Figures in **BOLD** are statistically significantly different from zero

In many ways the omissions rate is a more meaningful statistic than the net under-count rate because in the net undercount calculation, omissions can be cancelled out by erroneous inclusions or double counting. A net undercount of zero could be the result of no one missed and no one double counted, or for example, 10% were missed, and 10% double counted.

Table 12.3 shows omissions rates by Race, Hispanic Origin and Tenure. For every group examined in Table 12.3 the omissions rates for the population living in rental housing units are higher than those for the population living in owner-occupied hous-ing units. Like net undercounts, Black Alone or in Combination, Native Hawaiian or Pacific Islander Alone or in Combination, and Hispanic renters have the highest omissions rates among the groups examined here (10% or more), indicating that these groups are particularly vulnerable to being missed in the Census.

Some of the more detailed categories of the population living in rental housing units show extremely high omissions rates. Detailed data from the 2010 Census Coverage Measurement program (U.S. Census Bureau 2012, Table D), for example, found Black Alone or in Combination males age 30–49 living in rental housing units had an omissions rate of 19.7% and Black Alone or in Combination males age 18–29 living in rental housing units had an omissions rate of 16.9%. For Hispanic Origin males age 18–29 the omissions rate was 16.1% and for American Indian and Alaskan Native Alone or in Combination males age 30–39 the omissions rate was 12.5%. Word (1997) also shows the powerful effect of tenure on self-response rates in the 1990 Census.

Table 12.3 2010 census omissions rates by race, Hispanic Origin, and tenure

	Percent omitted	
	Population living in owner-occupied housing units	Population living in rental housing units
Total	3.7	8.5
Non-Hispanic White Alone	3.0	6.4
Black Alone or in Combination	7.2	11.0
American Indian and Alaskan Natives Alone or in Combination	6.6	8.8
Asian Alone or in Combination	4.1	7.4
Native Hawaiian or Pacific Islander Alone or in Combination	5.6	10.1
Hispanic Origin	5.0	10.4

Source U.S. Bureau of the Census, 2012, DSSD 2010 CENSUS COVERAGE MEASUREMENT MEMORANDUM SERIES 2012, 2010-E-51, Tables A and B
Note The directionality of undercounts and overcounts have been reversed from the original publication in order to keep them consistent within this publication In other words, a negative sign implies an undercount
Figures in **bold** are statistically significantly different from zero

12.5 Net Coverage Rates Over Time by Tenure

Table 12.4 shows net undercount rates by tenure for the 1990, 2000 and 2010 Censuses. In every Census since 1990, there was a net undercount for the population living in rental housing units and a net overcount for the population living in owner-occupied housing units (there was a small net overcount for owners in 1990 but it rounds to zero). The net undercount rate for renters fell from 4.5% in 1990 to 1.1% in 2000, but it did not change between 2000 and 2010. The Census coverage gap between the population living in owner-occupied units and those living in renter-occupied units went from 4.5 percentage points in 1990 to 1.7 percentage points in 2010.

Table 12.4 Net undercount rates by tenure: 1990, 2000 and 2010

	2010	2000	1990
Population in owner-occupied housing units	**0.6**	**1.3**	0
Population in rental housing units	**−1.1**	**−1.1**	**−4.5**

Source U.S. Bureau of the Census, 2012, DSSD 2010 CENSUS COVERAGE MEASUREMENT MEMORANDUM SERIES 2012, 2010-G-1 Table 10
A negative sign reflects a net undercount. The signs here are reversed from the source report in order to keep directionality consistent within this publication
Figures in **bold** are statistically significantly different from zero

Table 12.5 Poverty and median household income by tenure: 2015 ACS

	Family poverty rate (percent in poverty)	Median household income
Population in owner-occupied units	7	$73,127
Population in rental units	27	$37,264

Source Author's analysis of U.S. Census Bureau's 2016 American Community Survey PUMS file on IPUMS system at the University of Minnesota and ACS Table B25119 Downloaded from American Factfinder

There does not seem to be any linear temporal pattern to the net coverage rates of people living in owner-occupied housing units. The net undercount for the population living in owner-occupied housing units went from zero in 1990 to a net overcount of 1.3% in 2000 but then showed a net overcount rate of 0.6% in 2010.

12.6   Tenure and Socioeconomic Status

In some ways the coverage differentials reflected by tenure are symptomatic of socioeconomic differences and tenure is sometimes used as a proxy for differences by socioeconomic status. Table 12.5 shows the family poverty rates and the median household income for renters and homeowners. The family poverty rate for the those living in renter-occupied units (27%) is almost four times that of the population living in owner-occupied units (7%) and the median household income renters ($37,264) is about half that of owners ($73,127). The results here underscore the connection between socio-economic status and Census coverage.

Another reason renters tend to be covered less well than owners is that many rental units are in multi-unit structures and living in a multi-unit structure is associated with being difficult to enumerate. Among people in single family housing units (including mobile homes) only 20% are in rental units compared to 78% in multifamily structure being in rental units. Also, the population living in rental housing units are more difficult to enumerate because they are more likely to move. The 2016 ACS (U.S Census Bureau 2018b) shows 25% of people in rental units moved in the previous 12 months compared to 8% in owner-occupied units.

12.7   Summary

The population living in rental housing units has a net undercount while those living in owner-occupied housing units are overcounted. In addition, the population in rental housing units have higher omissions rates than the population living in owner-occupied housing units. For every racial group examined here renters tend to be covered less well than owners.

References

U.S. Census Bureau. (2012). *2010 components of census coverage for race groups and Hispanic origin by age, sex, and tenure in the United States*. DSSD 2010 census coverage measurement memorandum series #2010-E-51. U.S. Census Bureau, Washington, DC.

U.S. Census Bureau. (2018a). *Downloaded from American Factfinder, Table B25008)*.

U.S Census Bureau. (2018b). *Downloaded from American Factfinder, Table B07013)*.

Word, D. L. (1997). *Who responds/who doesn't? Analyzing variation in mail response rates during the 1990 census*. Population Division Working Paper No. 19, U.S. Census Bureau, Washington DC.

Open Access This chapter is licensed under the terms of the Creative Commons Attribution 4.0 International License (http://creativecommons.org/licenses/by/4.0/), which permits use, sharing, adaptation, distribution and reproduction in any medium or format, as long as you give appropriate credit to the original author(s) and the source, provide a link to the Creative Commons license and indicate if changes were made.

The images or other third party material in this chapter are included in the chapter's Creative Commons license, unless indicated otherwise in a credit line to the material. If material is not included in the chapter's Creative Commons license and your intended use is not permitted by statutory regulation or exceeds the permitted use, you will need to obtain permission directly from the copyright holder.

Chapter 13
Potential Explanations for Why People Are Missed in the U.S. Census

Abstract Knowing the characteristic of people most likely to be missed in the Census is not the same as knowing why they are missed. In this Chapter information is provided on several of the leading ideas about why people are missed in the Census along with data related to many of the ideas. The topic is first approached from broad theoretical perspective then more detailed reasons are examined. The chapter draws heavily on the literature in survey research methodology.

13.1 Introduction

Most of this book has focused on the characteristics of people and groups that have high census net undercount and omissions rates. But knowing the characteristics of people who are missed is different from knowing why they are missed. This Chapter explores several possible reasons why people are missed in the U.S. Census and examines statistical data related to many of the ideas. Some of the ideas examined here reflect broad factors (like the characteristics of households and living arrangements) and some reflect narrow factors (like the imputation algorithm used by the Census Bureau).

It is widely believed that there is no single reason why people are missed in the Census. With respect to differential undercounts of racial and ethnic groups, de la Puente (1993, p. 2) captured the dominant perspective more than 20 years ago,

> Empirically and in the aggregate, there is no single reason why a disproportionate number of ethnic and racial minorities were not counted by the 1990 census. Rather, there are a constellation of factors that interact and contribute to the differential undercount.

The quote above focuses on undercounts by race and ethnicity, but the lack of a single predominant reason for people being missed in the Census goes beyond race and ethnic groups. With respect to the high net undercount of young children, the Census Bureau Task Force on the Undercount of Young Children (U.S. Census Bureau 2014) concluded, "The task force is convinced that there is no single cause for this undercount".

© The Author(s) 2019   123
W. P. O'Hare, *Differential Undercounts in the U.S. Census*,
SpringerBriefs in Population Studies, https://doi.org/10.1007/978-3-030-10973-8_13

Despite several reports that address the issue of why people are missed and/or barriers to enumeration in the Census (U.S. Census Bureau 1992; de la Puente 1993; Martin and de la Puente 1993; Simpson and Middleton 1997; Schwede and Terry 2013; West and Fein 1990) there is no consensus framework for examining this issue. Because the Decennial Census is a large and complex operation, there is widespread belief that people are missed in the Census for many different reasons, and researchers approach census omissions from many different disciplines and perspectives, it is not surprising that there is no consensus on the best framework to use for understanding this problem.

13.2   What Is an Omission?

Determining what is an omissions in the Census is not as easy as it may seem. For example, in the 2010 Census there were 6 million whole person imputations (U.S. Census Bureau 2012a, Table 6). People are imputed into the Census count if they don't self-respond and there is good evidence they exist (for example, a housing unit that looks occupied), but the Census enumerator can never find anyone at home. If a person does not respond to the Census but data are imputed to represent them in the census count is that an omission?

Also, some households and persons are included in the Census by proxy respondents. If a Census enumerator is unable to contact residents of a household after several attempts, the enumerator may seek data on the household from someone like a neighbor or a landlord. These are referred to as proxy respondents. Should people included in the census count by proxy respondents be called omissions?

It is also important to make a distinction between people who are totally missed in the Census and those who are included but misclassified. For example, if a Black person is included in the Census, but has their race coded as White (perhaps because that characteristic was imputed) that would not impact overall omissions, but it would impact the omissions for Blacks and Whites.

13.3   Broad Ideas About Why People Are Missed in the Census

Many of the ideas or theories about why people are missed in the Census are linked to broader questions of non-responsiveness in surveys (National Research Council 2013; Groves and Couper 1998). Several leading theories for understanding omissions in surveys and censuses from a theoretical perspective, including social capital theory, leverage-saliency theory, social exchange theory, are discussed in a report by the National Research Council (National Research Council 2013, pp. 33–36).

Table 13.1 Why barriers leads to being missed in the census

Barrier	How barriers leads to being missed in the census
Complex households	In complex households, more people are likely to be unrelated to the householder. Respondents are often uncertain whether to include someone who is not directly related to the householder. In addition respondents may be confused about concepts such as household and family and staying temporarily versus living at the census address
Irregular housing units	Irregular housing units may not get on the Master Address File and therefore not receive a Census questionnaire
Lack or cooperation/trust	People who don't want to cooperate with the Census Bureau, because they don't trust the Census Bureau to treat their data confidentially or for other reasons, are less like to respond to the Census
Communication/language	Respondent who are not comfortable in English may be more reticent to respond to the Census. For some people, dealing with a language other than the one they are comfortable with, is just one more hurdle to participating in the Census
Renters	In general renters have less attachment to their community than home-owners and therefore may be less likely to engage in this civic activity so they are less likely to respond to the Census. Renters are also more mobile
Socio-economic status	People with lower socio-economic characteristics like poverty, unemployment and lower levels of education, are pressed by issues other than the Census. They are less likely to see the salience of returning a census questionnaire. Some may not even receive a questionnaire (related to other barriers)
Residential mobility	People who are highly mobile around Census day may be moving between housing units and not receive a census questionnaire. Renters are also less committed to their community and may not see the salience or the benefits or returning a census questionnaire
Non-city style/non-tradition addresses	These housing units are less likely to be on the Master Address File and therefore may not receive a Census questionnaire

Note This table draws heavily from Appendix 1, in Robinson et al. (2007)

A report by Robinson et al. (2007) lists several barriers to census enumeration and in Appendix 1 of the report they explain how each barrier is tied to the likelihood of being missed in the Census. Table 13.1 lists the barriers offered by Robinson et al. (2007) and a short explanation of how each barrier could results in someone not be counted in the Census.

Other researchers have provided similar ideas about why people are missed in the Census (West and Fein 1990; de la Punte 1993; Schwede et al. 2015).

13.4 People Missed in the Census Due to Failure of Steps in the Data Collection Process

Another approach to understanding why people are missed in the Census is offered by researchers who examine the steps in the data collection process and try to understand which step failed and why (Beimer et al. 1991). Many researchers start by decomposing Census coverage along the lines of Olson (2009) who posits that any omissions in the Census-taking process must come from a failure of one of three steps below;

- Failure to enumerate housing unit,
- Failure to get a complete and accurate roster of household members,
- Failure to get information for a person on the roster.

To reconfigure these three reasons for omissions, the first means the household did not get enumerated, and the second and third, means the household was enumerated, but not everyone in the household who should have been included, was included in the Census count.

13.5 Missing Households

Table 13.2 shows 4.2 million housing units were missed in the 2010 Census. Of the 4.2 million housing units missed, 2.2 million were occupied. This estimate is based on comparing the housing units found in the Post-Enumeration Survey to those found in the Census (U.S. Census Bureau 2012a). It is easy to understand why someone living in an overlooked housing unit might be missed in the Census. It is unlikely that the omitted housing units received a census questionnaire and therefore it is unlikely that occupants of those housing units were included in the 2010 Census.

The number of persons in the missed households is not directly available from the Census Bureau, but if the 2.2 million occupied households that were missed had the average number of persons per household in the 2010 Census (2.58) (U.S. Census

Table 13.2 Housing units omitted in the 2010 census

	Census count of housing units (in millions	Number omitted (in millions)	Percent omitted
U.S. total	131.7	4.2	3.2
Occupied	116.7	2.2	1.9
Owner-occupied	76.0	1.2	1.6
Renter-occupied	40.7	1.0	2.5
Vacant	15.0	2.0	13.2

Source U.S. Census Bureau (2012a), Table 3

Bureau 2018a) the missed households would account for 5.7 million missed persons out of a total of almost 16 million omissions in the 2010 Census (U.S. Census Bureau 2012b, Table 2). Missed housing units appears to be a big reason for the high number of omissions in the 2010 Census.

There are several reasons why housing units may be missed in the Census. In some cases, homeowners turn a basement or a garage into a separate housing unit so what appears to be a single-family residence is a multi-unit structure and the new housing units are easy to miss. Mahler (1993) indicates landlords sometime create illegal units in multi-unit structures and they are reluctant to reveal the existence of these units.

In many rural areas housing units may be far off the road and hidden from view and so they are not included in the Master Address File. For example, de la Puente (1993, p. 15) states,

> In rural areas, unlike in urban areas, unmarked and/or hidden roads and mismatches between mail delivery address of housing units and the actual physical location of the housing unit are conditions associated with the omissions of housing units from the Census.

Also, in both urban and rural settings Census Bureau canvassers must determine if a housing unit is inhabitable. If a building is deemed uninhabitable it will not receive a Census questionnaire. Making a judgment about the inhabitability of a structure is difficult, despite guidance provided by the Census Bureau; but just as important, some structures that are deemed uninhabitable may actually have people living there. Kissam (2017) and Fein (1989) provide detailed descriptions of some of the many ways housing units are likely to be missed in the Census.

13.6 People Omitted on Census Questionnaires that Are Returned

In addition to people who are not counted because their whole household was missed, some people may be missed because they were left off Census questionnaires. People may be omitted because they are left off a questionnaire that was mailed back, missed by an enumerator during the Non-Response Followup operation or possibly missed in another way such as an incorrect proxy response, a data processing error, or incorrect imputation.

Most omissions can be classified as the result of either confusion or concealment. Confusion can lead to a missed person if the respondent is confused about whether a given person should be thought of as part of the household or family and included in the census questionnaire. Confusion can also lead to a person being missed if the respondent is unsure if a certain type of person (like a young child or noncitizen) is supposed to be included in the Census.

Other people may be left off Census questionnaires because the respondent wants to conceal a person or whole household from the authorities. There are many reasons why a person might be concealed. For example, some people fear the federal government and they see the Census Bureau as just another federal government agency. Growing fear of the federal government in many communities is likely to be a big factor in the 2020 Census.

13.7 People Omitted in the Census Because of Confusion

The "usual place of residence" is a key concept used by the Census, but Martin (1999, 2007) argues that concept is not always clear to respondents and attachment to a single household may not apply to some people. Martin talks about the concept of "residential ambiguity" which reflects uncertainty about whether an individual belongs to a housing unit or household. Moreover, most of the rules respondents usually use to determine who they think lives in their household (economic contributions, doing household chores, receiving mail at the address) are not the rules used in the Census, so respondents may be confused about the concepts of families and households used in the Census. According to West and Robinson (1999, p. 10),

> The Census rules of residence instruct that the person in whose name the house or apartment is owned, being bought or rented be listed as person 1 on the form. The respondent is then asked to identify members of the household in relation to person 1. This often contradicts the respondent's notion of family or household.

It is easy to understand how someone who is only marginally attached to a housing unit may be missed in the Census. Whoever is filling out the Census questionnaire for a household may think the marginally attached person does not really live in the housing unit, so they are not entered on the Census roster. It is also feasible that the person filling out the census questionnaire may think someone who does not live at a housing unit continuously, for example, children in joint custody or someone who travels on business regularly, is being counted elsewhere. In reporting on confusion on the part of respondents in the 2010 Census, Schwede and Terry (2013, p. 89) concluded, "Additionally, the situation of mobility of people cycling between housing units and trying to determine from time spent in each where they should be counted was a reason for inconsistencies".

The problems with residence rules have been noted before. According to The National Research Council (2004, p. 153),

> The current definition of residence rules is confusing both to field enumerators and to residents. Difficulties arise for people with multiple residences, including those with movement patterns that are primarily within a week, or those that move seasonally. Such movement patterns are typically true of retirees, those involved in joint custody, arrangements, those with weekend homes in the country, students away at college during the school year, and people temporarily overseas.

One common situation related to confusion about the occupants of a housing unit is people living in complex households (Schwede 2018). For example, West and Robinson (1999, p. 9) describe one situation that may lead to a child being missed in the Census.

A child who resides in a diverse household structure and in a unique living arrangement among multiple nuclear families…Unusual living arrangements involving people that make it difficult for the respondent to roster the household correctly on the Census form, e.g. presence of multiple nuclear families, unrelated people or step people of the respondent.

Complex households often involve the presence of subfamilies in a household and that can make correct rostering of household members more complicated. One official at the Census Bureau "…noted that she was aware of instances with multiple families, for example, where the household respondent did not include people in the second family" (cited in U.S. Census Bureau 2014, p. 16). Martin (2007) contends people more remotely linked to the person filling out the census questionnaire, are more likely to be missed.

A report from the U.S. Census Bureau (2016) indicates that many young mothers were not included in the 2010 American Community Survey (ACS) and therefore they argue many were probably unreported in the Census. Many of the young mothers missed in the ACS were probably single mothers with their child(ren) living with the parent(s) of the mother or with some other householder. It is feasible that the respondent believes the young mother and her children are only living with them temporarily and so does not enter them on the Census questionnaire roster. It is also easy to imagine a grandmother filling out the Census questionnaire may think of the young grandchild in her household as part of her daughters' family rather than part of her family.

A rapidly growing type of living arrangement for people (percentage-wise) is cohabitating households. Since cohabiting couples reflect living arrangements that are relatively unstable (compared to married-couple families) and the relationships among adults and children are different from those in a nuclear family, it would not be surprising if a disproportionately high share of people in cohabiting households were not being reported in the Census.

Newborns may be particularly likely to be living in complex households A recent report from the Census Bureau (Monte and Ellis 2014, p. 2) found "more than one in five women with a birth in the past 12 months reported at the time of the survey that they were living in someone else's home". In another analysis (Gooding 2008) shows that 13% of mothers were not co-residing with their biological child under age 1 and rates are higher for Blacks and Hispanics where the net undercount of people is also higher. This may help explain the high omissions rate for young children (O'Hare 2015).

13.8 Large and Complex Households

People living in large households may be missed for many different reasons. First, large households are often complex households and people may be left off the questionnaire for reasons described in the previous section. Martin (1999, 2007) shows that being closely related to the respondent in a survey greatly decreases ambiguity related to residential attachment and complex households are more likely to have occupants who are not closely linked to the householder.

Examination of survey data taken prior to the 2010 Census shows that having more than two people in the household lowers the likelihood of respondents saying they will participate in the 2010 Census independent of other factors (Walejko et al. 2011). Data from the 2010 Census show that 84% of 2-person households mailed back their Census questionnaire compared to only 72% of 7-person households (Letourneau 2012).

Larger households are likely to contain children and child care demands may interfere with completing the Census questionnaire. This is a point made by Hillygus and colleagues (2006, p. 103) who note,

> Respondents who are married with children have a lower mail-back rate (83 percent) than those who are married without children (90 percent), suggesting that the time demands of child care work against taking on this particular duty.

In a poll conducted just prior to the 2010 Census (Pew Research Center 2010) about a third of those who said there were not planning to participate in the 2010 Census cited "too busy or not enough time" as the reason.

13.9 Confusion About What Types of People Should Be Included in the Census

Another reason people are missed in the Census is because respondents may believe the Census Bureau does not want some categories of people included in the Census. In a series of short surveys by the Census Bureau (Nichols et al. 2014a, b, c) respondents were asked, "What information do you think the Census typically collects every 10 years?" and were offered several choices. The percentage who thought the Census Bureau collects "Names of children living at your address" was 7–9 percentage points lower than the percentage who thought the Census Bureau collects, "Names of adults living at your address". While this question asks about names rather than about information on individuals, it suggests that some people think the Census does not request information on children.

A recent report (Vargas 2018) based on a nationwide survey of Latinos conducted by the National Association of Latino Elected Officials found that 15% of respondents who had a child under age 5 in the home would not count them in the Census or do not know if they would count them. These attitudes help explain why young children have a higher omissions rate than any other age group in the 2010 Census (O'Hare 2015). Also, in their qualitative study of 2010 Census respondents Schwede and Terry (2013) indicated many respondents do not believe the Census Bureau (or the federal government) wants children included in the Census count.

Young children are not the only group some people think are not supposed to be included in the Census. Data from the Census Barriers Attitudes and Motivators Survey done in 2008, indicates only 76% of the people interviewed think the Census Bureau wants non-citizens included in the Census (U.S. Census Bureau 2009, p. 88). In a similar study in 2018 (U.S. Census Bureau 2018b) only 55% of respondents were sure that noncitizens were supposed to be counted in the Census.

13.10 People Deliberately Concealed

There is evidence to support the idea that some people may be left off Census questionnaires on purpose. In some situations, there may be an effort to conceal an entire household, and in other situations the effort may be focused on concealing selected individual(s) within a household.

Some people may also be purposively left off census questionnaires based on rules about the housing unit where they are living. People could be left off because there are too many people living in the housing unit relative to rules about maximum capacity. If a housing unit is only supposed to have four people, by rule, but there are five people who regularly live there, one may be left of the census questionnaire because the respondent fears reporting five people in the unit might jeopardize continued occupancy.

Some people may be left off a census questionnaire because of the types of people who are allowed to live in a housing unit. For example, the report of the Census Bureau Task Force on the Undercount of People (U.S. Census Bureau 2014, p. 16) states one of the reasons people are left off forms is, "Respondents deliberately not mentioning kids for fear of some reprisals or bad outcomes from landlords, immigration agencies, social service agencies, etc". West and Robinson (1999, p. 7) also conclude,

> Listing some members of the households may have other negative consequences. For example, a respondent may fear that disclosure of certain members of the household will affect eligibility for social services, that people illegally in the country will be deported, or that the whereabouts of a child in hiding from a custodial parent will be detected.

Pitkin and Park (2005) also mention "systematic concealment" as a potential reason children are undercounted in the Census.

Concealment may be driven by lack of trust in the Census Bureau and/or the federal government. Despite assurances from the about the confidentiality of responses to the Census (Congressional Research Service 2018), many people believe that data given to the Census Bureau may be shared with other government agencies. A poll (Pew Research Center 2010, p. 5) conducted just prior to the 2010 Census indicated that among those who said they were not planning to participate in the 2010 Census,

18% cited distrust of government and 8% cited privacy concerns. In the 2010 Census Barriers Attitudes and Motivators study (U.S. Census Bureau 2011) researchers classified 10% of the population as cynical and 14% as suspicious of the Census Bureau.

Census Bureau staff report growing number of respondents are refusing to cooperate because of fear (Meyers and Goerman 2018; U.S. Census Bureau 2017a, b). There is growing distrust of the federal government and for most people the Census Bureau is seen as another branch of the federal government. Abowd (2018) indicates that fear among undocumented population is also likely to be have an impact on many of the people who are in the country legally because they live in a household where undocumented people are living. One recent report concludes (Alsan and Yang 2018, p. i). "Though not at personal risk of deportation, Hispanic citizens may fear their participation could expose non-citizens in their network to immigration authorities". This report was focused on participation in safety net programs, but I think it applies to the Census as well.

Such distrust affects certain populations more than others. Immigrants are one population where concealment may be common because of fear of the federal government. The 2016 ACS (Table S0501) shows there are 43.7 million foreign-born residents in the U.S. and 22 million noncitizens. The Pew Research Center (2018) reports that there were about 11 million unauthorized immigrants in the U.S. in 2015. The estimate from the Pew Research Center is similar to estimate for January 2014 from Homeland Security (2018) of between 10.8 million and 12.1 million. This impacts not only noncitizens and undocumented immigrants but the people living in a household with them. There are 6.4 million children living in a household with at least on undocumented immigrant. Abowd (2018, p. 6) states, "From the 2016 ACS we estimate that 9.8% of all households contain at least one noncitizen". Almost all children less than 5 years old are citizens but 20% of children in this age range live with at least one noncitizen (Population Reference Bureau 2018).

13.11   Barriers Posed by Questionnaire Design

There are couple of aspects of the design of the Census questionnaire that may contribute to omissions in the Census. Both are related to rostering or listing all the people in the housing unit.

The first step in the Census-taking process is getting a respondent to list all the people in the household. This is called rostering. There is a lot of evidence indicating that the way rostering is done can impact who is included in the roster. For example, West and Robinson (1999, p. 6) conclude, "Coverage errors are likely to occur because the respondent has difficulty rostering his or her household". A recent paper by Battle and Bielick (2014) suggests that the inclusion of children may be particularly sensitive to the way rostering is done. An experiment by Tourangeau and colleagues (1997) found substantial variation in persons who were deemed to live at a given address with differing rostering instructions. Other researchers (Lin et al. 2004; Waller and Jones 2014) found rostering instructions very important in determining who is included in a given household. As stated earlier, the rules for whether someone should be listed as part of the household are not always clear to respondents. One perspective on the mismatch between the Census questionnaire and changing American families is provided by Jacobsen (2017) who contends, the Census Bureau's data collection methods have not kept pace with the rapidly changing American family.

It should be recognized that alternative ways of rostering households often have tradeoffs. Some rostering methods result in higher respondent burdens and rostering methods that impose a higher burden on respondents are likely to reduce response rates.

On the Mailout/Mailback Census questionnaire that was used in the 2010 Census there was only room for complete demographic information for six people in the household. There is limited room for the names and a few characteristics for the 7th through the 12th person. If more than 12 people lived in the household, the Census Bureau had to follow up to get information for these people. For people living in the largest households (13 or more people) they may be missed because there is not enough room on the questionnaire for everyone in the household to be listed and follow up failed.

When incorrect or incomplete information is provided on the 2010 census questionnaire, the Census Bureau followed up with the household to get complete information for persons. But followup was often problematic, and the Census Bureau was only able to contact a little more than half the people in the 2010 Census followup operation (U.S. Census Bureau 2012d). The fact that followup was only done by telephone (not face to face follow up) also hampered data collection. Flaws in the followup operation may result in an omission. Heavy use of the internet for data collection in the 2020 Census may help lessen this issue because followup will be immediate on the internet.

13.12   People Missed Because of Estimation and Processing Errors

In addition to whole households being missed and people being left off census questionnaires, some people may not be reflected in the Census count because of processing errors in the Census operations. In particular, the final census counts include many people who are imputed and a large number who are included by proxy responses.

Imputations take several forms. The simplest form is item imputation. If a respondent leaves a census question (an item) blank, say race, the Census Bureau uses data from the person, the household, and the neighborhood to impute a race for the person. This might impact omissions figures for certain groups. For example, if a person who is really Black had their race imputed as White, the omissions for Blacks would look higher and the omissions for Whites would look lower.

A bigger issue in terms of omissions is whole-person imputation. In the 2010 Census there were about 6 million persons imputed (U.S. Census Bureau 2012c, Table 9). This also takes several forms. If the Census Bureau enumerator has failed to find someone at home after repeated attempts, they may ask a proxy respondent such as a landlord of neighbor about the people living in the housing unit. It is not difficult to imagine a neighbor or landlord saying there are three people living in the housing unit when in fact they are four which results in an omission. Proxy responses provide low-quality data compared to self-reporting. The U.S. Census Bureau (2012b, Table 12) reports that 93% of responses from household members were correct compared to only 70% of those from proxies. The imputation methods of the Census Bureau have been refined over many censuses, but still may result in errors.

If after repeated visits when the Census Bureau enumerator found no one was at home, and no proxy respondent can be found, and the occupancy status of a housing unit is unclear, the Census Bureau imputes an occupancy status (occupied or vacant). If the imputed occupancy status is "occupied" the Census Bureau, then imputes the number and characteristics of people for the housing unit. It is easy to imagine a housing unit that really has four people living there, only gets three people imputed, and thus one person is omitted. It is also possible that several squatters may be living in a building that is imputed as vacant, so they are all missed in the census.

13.13 Summary

Several potential explanations for census omissions were examined and statistical data were provided for most potential explanations. While there is more support for some potential explanations that for others, no single reason or theory seems completely compelling. Perhaps the most fundamental conclusion from the material reviewed in this Chapter is that there are many different reasons why people are missed in the Census.

Some people are missed because the housing unit where they live is not included in the Census and others are missed because they are not captured in the Census even though others in the housing unit where they live are. Other findings include:

- People may be missed because they are more likely to live in complex or non-traditional households where their status in the household is unclear.
- Some people are missed because respondents are confused about who should be included on their Census questionnaire.
- People may be missed because respondents want to conceal them from the government, in part, because of fear or reprisals or negative outcomes.
- Some aspects of the census-taking process (like the construction of the questionnaire) result in some people being missed.

References

Abowd, J. M. (2018). "Technical review of the department of justice request to add citizenship status question to the 2020 census," Memorandum for Wilbur L. Ross Jr., January 19, 2018.

Alsan, M., & Yang, C. (2018). *Fear and the safety net: evidence from secure communities.* Working Paper 24731, National Bureau of Economic Research, Cambridge, MA.

Battle, D., & Bielick, S. (2014). *Differences in coverage and nonresponse when using a full household enumeration screener versus a child-only screener in a 2013 national mail survey, paper presented at annual conference of the American Association of Public Opinion Research.* CA June: Anaheim.

Beimer, P. P., Groves, R. M., Lyberg, L. E., Mathiowetz, N. A., & Sudman, S. (1991) *Measurement errors in surveys.* Wiley.

Congressional Research Service. (2018). "Confidentiality provisions for the 2020 decennial census," memorandum to house committee on oversight and government reform, From Jennifer Williams, dated April 19 Washington, DC.

de la Puente, M. (1993). Using ethnography to explain why people are missed or erroneously included by the census: evidence from small area ethnographic studies. Washington, DC: U.S. Census Bureau.

Fein, D. J. (1989). The social sources of census omissions: Racial and ethnic differences in omissions rates in recent U.S. censuses. Dissertation in Department of Sociology, Princeton University, Princeton, NJ.

Gooding, G. E. (2008). Differences between coresident and non-coresident women with a recent birth. Annual Meeting of the American Sociological Society, Boston, MA, August.

Groves, R. M., & Couper, M. P., (1998). *Nonresponse in household interview surveys.* Wiley.

Hillygus, S. D., Nie, N. H., Prewitt, K., & Pals, H. (2006). *The hard count: The political and social challenges of census mobilization.* New York, NY: Russell Sage Foundation.

Homeland Security. (2018). *Estimates of the unauthorized immigration population residing in the United States, January 2014, Office of Immigration Statistics, Office of Strategy, Policy and Plans.* https://www.dhs.gov/sites/default/files/publications/Unauthorized%20Immigrant%20Population%20Estimates%20in%20the%20US%20January%202014_1.pdf.

Jacobsen, L. (2017). Discussion. *Journal of Official Statistics, 33*(4), 891–899.

Kassim, E., (2017). Differential undercount of mexican immigrant families in the U.S. Census. *Statistical Journal of the IAOS, 33*, 7979–816.

Letourneau, E. (2012). *Mail response/return rates assessment,* 2010 Census Planning Memorandum Series, No. 198, U.S. Census Bureau, Washington DC.

Lin, I. F., Schaeffer, N. C., & Seltzer, J. A. (2004). Divorced parents qualitative and quantitative reports of children's living arrangements. *Journal of Marriage and Family, 66,* 385–397.

Mahler, S. (1993). Alternative enumeration of undocumented Salvadorans on Long Island, Prepared under Joint Statistical Agreement 89-46 with Columbia University, Bureau of the Census, Washington DC.

Martin, E. (1999). Who knows who lives here? Within-household disagreements as a source of survey coverage error. *Public Opinion Quarterly, 63,* 220–236.

Martin, E. (2007). Strength of attachment: Survey coverage of people with tenuous ties to residences. *Demography, 44*(2), 437–440.

Martin, E., & de la Puente, M., (1993). *Research on sources of under coverage within households,* U.S. Census Bureau, Washington, DC.

Meyers, M., & Goerman, P. (2018). Respondents confidentiality concerns in multilingual pretesting studies and possible effects on response rates and data quality for the 2020 census. Paper delivered at the annual Conference of the American Association of Public Opinion Research, Denver CO., May 16–19, 2018.

Monte, L. M., & Ellis, R. R. (2014). Fertility of women in the United States: 2012, Population Characteristics, P20-575, U.S. Census Bureau, Washington DC.

National Research Council. (2013). Nonresponse in social science surveys. In R. Tourangeau & T.J. Plewes (Eds.). Washington, DC: National Academy Press.

National Research Council. (2004). *Reengineering the 2010 census: Risks and challenges*. In D.L. Cork, M.L. Cohen, & B.F. King (Eds.), Washington, DC: National Academy Press.

Nichols, E., King, R., & Childs, J., (2014a). Small-scale testing pilot test results: Testing email and address collection screens and Census opinion questions using a nonprobability panel. Internal memorandum to Burton Reist. Census Bureau (March 27).

Nichols, E., King, R., & Childs, J., (2014b). 2014 march small-scale testing pilot test results: Testing email subject lines, email formats, address collection screens and Census opinion questions using a nonprobability panel. Internal memorandum to Burton Reist. Census Bureau (May 27).

Nichols, E., King, R., and Childs, J., (2014c). May 2104 small-scale testing results: Testing email subject lines, email formats, address collection screens and Census opinion questions using a nonprobability panel. Internal memorandum to Burton Reist. U.S. Census Bureau. (September 9).

O'Hare, W. P., (2015) The undercount of young children in the U.S. decennial census. Springer Publishers.

Olson, D. B. (2009). *A three-phase model of census capture*. Paper presented at the Joint Statistical Meetings.

Pew Research Center. (2010). Age, education, ethnic and partisan gaps: Most view census positively, but some have doubts. New Release January 20, Washington DC.

Pew Research Center. (2018). *5 facts about illegal immigration in the U.S., fact tank downloaded*, July 15, 2018. http://www.pewresearch.org/fact-tank/2017/04/27/5-facts-about-illegal-immigration-in-the-u-s/.

Pitkin, J., & Park, J. (2005). *The gap between births and census counts of people born in California: undercount or transnational movement?* Paper presented at the Population Association of America Conference, Philadelphia PA, March.

Population Reference Bureau (2018). Citizenship question risks a 2020 census undercount in every state, especially among children. Population Reference Bureau, Washington DC.

Robinson, J. C., & Bruce, A. (2007). The planning database: decennial census data for historical, real-time, and prospective analysis, paper presented at Joint Statistical Meetings, 2007, Salt Lake City.

Schwede, L. (2018). Linkages among the rise in complex households, the undercount of young children, and race/ethnicity: where we have been and where we might go for the 2020 census. Poster Presented at the 2018 Annual Conference of the Population Association of America, Denver, CO, April

Schwede, L., & Terry, R., (2013). *Comparative ethnographic studies of enumeration methods and coverage across race and ethnic groups*, 2010 Census Program for Evaluations and Experiments, U.S. Census Bureau, Washington, DC.

Schwede, L., Terry, R., & Hunter, J. (2015). Ethnographic evaluations on coverage of Hard-to-Count minority in the US decennial censuses. In R. Tourangeau, B. Edwards, T. P. Johnson, K. M. Wolter, & N. Bates (Eds.), *Hard-to-survey populations* (pp. 293–315). Cambridge, England: Cambridge University Press.

Simpson, L., & Middleton, E., (1997). *Who is missed by a national census? A review of empirical results from Australia, Britain, Canada, and the USA*. The Cathie Marsh Centre for Census and Survey Research, University of Manchester UK.

Tourangeau, R., Shapiro, G., Kearney, A., & Ernst, L. (1997). Who lives here? Survey undercoverage and household roster question. *Journal of Official Statistics, 13*(1), 1–18.

U.S. Census Bureau. (1992). *"Barriers Paper," Memorandum for Susan M. Miskura from LaVerne V. Collins*, October 29. Washington, DC: U.S. Census Bureau.

U.S. Census Bureau. (2009). *Census barriers, attitudes, dan motivators survey*. C2PO 2010 Census Integrated Community Research Memorandum Series, No. 11, May 18. Washington, DC: U.S. Census Bureau.

U.S. Census Bureau. (2011). *Census barriers, attitudes, dan motivators survey II final report*. 2010 Census Planning Memorandum Series, No. 205, June 26. Washington, DC: U.S. Census Bureau.

U.S. Census Bureau. (2012a). *2010 census coverage measurement estimation report: summary of estimates of coverage for housing units in the United States*. DSSD 2010 census coverage measurement memorandum series #2010-G-02, May 22, 2012. Washington, DC: U.S. Census Bureau.

U.S. Census Bureau. (2012b). *2010 census coverage measurement estimation report: Components of census coverage for the household population in the United States. 2010 Census coverage measurement memorandum series #2010-G-04*. Washington, DC: U.S. Census Bureau.

U.S. Census Bureau. (2012c). *2010 census coverage measurement estimation report: Summary of estimates of coverage for persons in the United States*. DSSD 2010 CENSUS COVERAGE MEASUREMNET MEMORANDUM SERIEs #2010-G-01. Washington, DC: U.S. Census Bureau.

U.S. Census Bureau. (2012d). *2010 census effectiveness of unduplication evaluation report*. 2010 Census Planning Memorandum Series, No. 244,Washington, DC: U.S. Census Bureau, October.

U.S. Census Bureau. (2014). *Final task force report: Task force on the undercount of young children*. Memorandum for Frank A. Vitrano. Washington, DC: U.S. Census Bureau. February 2.

U.S. Census Bureau. (2016). *2020 census research and testing, investigating the 2010 undercount of young children—Examining the coverage of young mothers*. Washington, DC, U.S: Census Bureau,

U.S. Census Bureau. (2017a). *"Respondent confidentiality concerns," Memorandum for Associate Directorate for Research and Methodology (ADRM) From Center for Survey Measurement (SCM)*, September 20, 2017.

U.S. Census Bureau. (2017b). *Respondent confidentiality concerns and possible effects on response rates and data quality for the 2020 census*. Presentation by Mikelyn Meyers at the Census Bureau National Advisory Committee Meeting, November 2, 2017.

U.S. Census Bureau. (2018a). *Downloaded from American Factfinder Table DP-1*.

U.S. Census Bureau. (2018b). *2020 Census barriers, attitudes, and motivators study (CBAMS) survey and focus groups: Key findings for NAC*. Presentation to the U.S. Census Bureau's National Advisory Group, November 1, 2108, slide 16.

Vargas, A. (2018). Slides from a webinar entitled census 2020: Research and messaging September 12. Available at https://drive.google.com/file/d/1QUga5owRyQyQY6IegH10i8ZkCrLalRkH September 12/view.

Walejko, G. K., Miller, P. V., & Bates, N. (2011). *Modeling intended 2010 census participation*. Paper delivered at the American Association of Public Opinion Research conference, Phoenix, AZ, May 30, 2011.

Waller, M. R., & Jones, M. R. (2014). Who is the residential parent? Understanding discrepancies in unmarried parents' reports. *Journal of Marriage and Family, 76*, 73–93.

West, K. K., & Fein, D. J. (1990). Census undercounts: An historical and contemporary sociological issue. *Sociological Inquiry, 60*(2), 127–141.

West, K., & Robinson, J. G., (1999). *What do we know about the undercount or children?* U.S. Census Bureau, Population Division working paper. Washington. DC: U.S. Census Bureau.

Open Access This chapter is licensed under the terms of the Creative Commons Attribution 4.0 International License (http://creativecommons.org/licenses/by/4.0/), which permits use, sharing, adaptation, distribution and reproduction in any medium or format, as long as you give appropriate credit to the original author(s) and the source, provide a link to the Creative Commons license and indicate if changes were made.

The images or other third party material in this chapter are included in the chapter's Creative Commons license, unless indicated otherwise in a credit line to the material. If material is not included in the chapter's Creative Commons license and your intended use is not permitted by statutory regulation or exceeds the permitted use, you will need to obtain permission directly from the copyright holder.

Chapter 14
Census Bureau Efforts to Eliminate Differential Undercounts

Abstract Over the past several decades the Census Bureau has engaged in many activities and programs aimed at reducing or eliminating differential undercounts. Several of the more prominent efforts to solve this problem are reviewed in this Chapter.

14.1 Introduction

Over the past several decades the Census Bureau has tried a number of approaches to reduce differential undercounts in the Census, but despite the best efforts of the Census Bureau, many differential undercounts have persisted. In this Chapter, a sample of steps the Census Bureau has taken to improve Census coverage and to reduce differential undercounts are presented. The steps involve things such as improved questionnaire design, field operations, Census promotion and outreach, as well as some changes in data processing.

The Census Bureau is certainly mindful of the differential undercount problem. Awareness of the problem is evidenced by a report following the 2000 Census (U.S. Census Bureau 2004a, p. 1) which states, "Censuses before 2000 have all been plagued by chronic undercount, and particularly by differential undercount of specific minority populations and other subgroups such as renters, males and children." The report goes on to say, "The pervasive nature of Decennial undercounts has strongly influenced Census design, including adding operations or programs specifically designed to improve coverage."

The programs initiated by the Census Bureau to address differential undercounts over the past several decades are far too numerous to discuss in detail here. Following the 1970, 1980, 1990 and 2000 Censuses (U.S. Census Bureau 1973, 1974, 1988, 1993, 2004b) the Census Bureau issued reports discussing the coverage improvement efforts for that Census in great detail. Unfortunately, there is no such report following the 2010 Census. These reports provide detailed descriptions of the efforts the Census Bureau has made to reduce differential undercounts. The U.S. General Accountability Office (2010a, b, c, 2017) has also provided a long stream of reports on the Census Bureau's attempts to improve the Census-taking process and the results.

© The Author(s) 2019

W. P. O'Hare, *Differential Undercounts in the U.S. Census*,
SpringerBriefs in Population Studies, https://doi.org/10.1007/978-3-030-10973-8_14

14.2 Undercount Adjustment

Probably the most visible attempt to remedy the problem of differential undercounts in the Census was the call to adjust Census figures to account for such undercounts (Choldin 1994; Darga 1999). During the late 1960s and early 1970s, there was a significant increase in popular understanding of Census undercounts, particularly in minority communities and among big city mayors. Some of the interest was stimulated by a paper by Hill (1975) which showed high net undercount rates and differentials for large cities. The increased interest in Census undercounts resulted in several changes to the 1980 Census procedures to try and eliminate differential undercounts (U.S. Census Bureau 1988). But there was little impact on the differential undercounts.

The focus during this period was largely on the differential undercount of Blacks and Whites or Non-Blacks. This was partly due to the availability of data on this differential, the large size of the differential, and the fact that Blacks were the largest racial or ethnic minority population at the time. The history of the Black/Non-Black differential is presented in Chap. 8.

The undercount adjustment movement peaked as the country approached the 2000 Census. The Census Bureau plans for the 2000 Census called for the official data to include an adjustment to rectify undercounts. In the late 1990s, as we moved toward the 2000 Census and Census Bureau plans for an undercount adjustment became clear, this issue took on a decidedly partisan political character. Because the Census Bureau's plan for adjustment involved extensive use of sampling, the term "sampling" mistakenly became synonymous with adjustment in many public and political debates. As Hannah (2001, p. 515) stated, "The recent controversy around the use of sampling methods in the U.S. Census illustrates some important political-geographic dimensions of our decisions regarding whether and how to be counted in surveys."

Given the fact that a disproportionate share of the people missed in the Census were racial minorities, and racial minorities tended to vote disproportionately for Democrats, adjustment was sometimes seen as a way to give more weight to Democratically leaning constituents. While producing adjusted figures probably would have helped Democrats more than Republicans, it is important to understand that the Census Bureau felt that the properly adjusted data would be more accurate. In other words, the decision to provide adjusted data for the 2000 Census was motivated by an interest in accuracy and not for political reasons.

One of the big impediments to producing official adjusted Census counts is the fact that adjusted figures would need to be produced very quickly after the Census count is completed. The Census Bureau must deliver the data used for political redistricting (Public Law 94-171 files) by March 31st of the year following the Census.

In the 2000 Census, the Census Bureau was prepared to provide adjusted Census data as the official figures but decided against releasing them at the last moment because the initial results appeared flawed and they were not confident in the results (U.S. Census Bureau 2002). A memo from Census Bureau Acting Director William

Barron (2001) to Secretary of Commerce Donald Evans stated, "As a member of the Executive Steering Committee for A.C.E. Policy (ESCAP) and as Acting Director, I concur with and approve the Committee's recommendation that unadjusted Census data be released as the Census Bureau's official redistricting data." The memo goes on to say, "The Committee reached this conclusion because it is unable, based on the data and other information currently available, to conclude that the adjusted data are more accurate for use in redistricting."

The Census Bureau later found an error in their procedures. According to U.S. Census Bureau (2002, p. i), "Evaluations of the March 2001 Accuracy and Coverage Evaluation (A.C.E) coverage estimates indicated the A.C.E. failed to detect a large number of erroneous enumerations." Following the 2000 Census, very little has been said about adjusting Census figures to account for undercounts.

14.3 Enhanced Outreach to Promote Participation in the Census

Efforts to educate the public about the importance of the Census and the need for everyone to participate have increased in recent decades. The Census Bureau (2016a, p. 1) states "for Censuses especially, publicity campaigns are a key component for success. An effective communication strategy delivers tailored messaging to audience segments using media and trusted voices."

Outreach efforts became bigger and more sophisticated after the 1990 Census. Bates (2017, p. 875) lists six interconnected elements of the Census Bureau's Social Media campaign of the past few decades;

1. Paid advertising
2. Earned media
3. Local and national partnerships
4. The Census website
5. Public relations
6. Census in schools.

A few of the most important programs listed above are discussed below.

14.3.1 Paid Advertising

One important intervention in the 2000 Census was the initiation of a paid advertising campaign. Prior to the 2000 Census, the Census Bureau had relied on free media like Public Services Announcements (PSAs). One problem with PSAs is that they were often run late at night or other times when viewership was low.

The paid advertising program was initiated in the 2000 Census and continued through the 2010 Census and is planned for the 2020 Census (U.S. Census Bureau

2017). In 2000 and 2010, the Census Bureau contracted with a private firm to oversee the paid advertising program. The paid advertising programs focused heavily on hard-to-count groups. The contract for the 2020 Census paid advertising program was awarded to the firm of Young and Rubican in August 2017 for $415 million (Media Post Agency Daily 2017). Unfortunately, some of the earliest activities under the contract had to be postponed because of the budget situation described in the next Chapter. Consequently, the paid advertising program has not gotten underway as quickly as the Bureau had hoped.

Paid ads appear to be effective in increasing response rates. Bates (2017, p. 876) states,

> In 1990, (the last Census to depend on a pro bono outreach campaign), the final mail response rate was projected at 70% but achieved only 65%. The 2000 Census (the first to use paid advertising), budgeted for a 61% response rate but achieved 67%. The 2010 Census (also with a paid campaign), also achieved a higher-than-projected mail response rate (projected was 64% with actual at 67%).

While it is likely that the paid advertising campaign increased the self-response rates in the 2000 and 2010 Census, it did not solve the differential undercount problem. For example, data in Chap. 5 indicate the net undercount of young children increased from 1990 to 2000 and again from 2000 to 2010, while the Census coverage rates for adults improved during this period.

14.3.2 Census Bureau Partnership Program

The Census Bureau Partnership Program, an attempt to get organizations outside of the Census Bureau involved promoting participation in the Census, began in the 2000 Census and expanded in the 2010 Census. The Partnership Program was an attempt to get "trusted voices" from hard-to-count communities to help deliver the message about the Census being important and safe (Olson et al. 2014). In the 2000 Census, there were about 140,000 Census partners and in 2010 about 255,000 (U.S. General Accountability Office (2010a, p. 13). Some of the Census partners were deeply involved in promoting the Census while others were partners in name only.

According to an evaluation of the 2010 Census National Partnership Program (U.S. Census Bureau 2012b, p. 9) here is a list of the types of groups and organizations that were sought for Partnerships with the Census Bureau,

1. Faith-based organizations
2. Labor unions
3. African-American population
4. Hispanic population
5. Asian/Pacific Islander population
6. American Indian or Alaskan Native population
7. Educational Institutions
8. Migrant organization

9. Recent Immigrant/Emerging populations
10. Government organizations
11. Federal government agencies
12. Disabled populations
13. Congressional organizations
14. Nonprofits organizations
15. Gay and lesbian populations.

With respect to the Partnership Program, the U.S. Census Bureau (2012b, p. 89) concluded, "The overall program has a measurable effect on increasing mailback rates in Hard-to-Count areas."

Census Complete Count Committees were another attempt started in the 2000 Census to expand outreach efforts around the Census and they are closely aligned with the Partnership Program. States and localities were encouraged to put together complete count committees to promote the Census in their jurisdiction. According to the Census Bureau (2018, p. 4), "Complete Count Committees (CCC) are volunteer committees established by tribal, state and local governments and community leaders or organizations to increase awareness and motivate residents to respond to the 2020 Census."

14.3.3 Census in Schools

The Census in School (CIS) program was started to reach households with children in the school system. The Census Bureau provides basic material on the Census to schools who agree to participate. There are two main goals in the Census in Schools Program. First, students learn about the importance of the Census in our system of government and second, they take home material that increase the chances their household will respond to the Census.

An evaluation of the Census in Schools program following the 2010 Census (U.S. Census Bureau 2012c, p. 107) concluded "Overall, teachers looking at the CIS materials during the focus group had a positive response to them and seemed glad to know about them for their future teaching." The Census in Schools program has been turned into an "evergreen" program called Statistics in Schools (U.S. Census Bureau 2016b).

In the 2000 Census, The Census in Schools program included a program for pre-schools but that was not included in the 2010 Census program. This is important because preschoolers (age 0–4) had a higher net undercount than any age group in the 2010 Census while the school age population (5–17) had a very low net undercount. As this book is being written, the plans for the Census in Schools program in the 2020 Census are not finalized.

14.4 Changes to the Census-Taking Process

Another approach to reduce differential undercounts relates to changes in the Census-taking operations. Several recent efforts are reviewed below.

Following the 2000 Census, the Decennial Census was changed to a short-form only Census in an effort to save money and increase participation. Prior to the 2010 Census, the Census involved a short form with only a handful of questions that were asked of everyone, and a long form which contained all the short-form questions as well as several dozen questions that were asked of a very large sample of households.

According to the Census Bureau (2003, p. 17) the mail return rate for the 2000 Census short form (76%) was 13 percentage points higher than the mail return rate for the long form (63%). In addition, the differential response rates between Blacks and Whites was a little higher for the long form than the short form. Among other things, it was felt that the change to a short-form only would increase participation and decrease undercount differentials. What was formerly the long-form Census has become the American Community Survey.

In the 2010 Census, the Census Bureau distributed bilingual questionnaires to areas with a concentration of Hispanic families for the first time. This measure seemed to increase response rates for Hispanics. After evaluating the 2010 Census experience with bilingual questionnaires, the Census Bureau (2011, p. v) concluded, "Further results suggest that the bilingual questionnaire provides substantial benefit to the areas that were targeted…".

In the 2010 Census, the Census Bureau sent replacement questionnaires to household in areas that had a low response rate in the 2000 Census in an effort to increase participation rates (U.S. Census Bureau 2012a). This was done two different ways. For areas that had the lowest response rates in the 2000 Census every household in the area received a replacement questionnaire. In other words, a few weeks after they received their first Census questionnaire, they received a second one even if they had already responded. This was referred to as "Blanketed areas" and there were about 53.7 million people in these areas. For areas that had a mid-range response rate in the 2000 Census, only nonresponding households received a replacement questionnaire. These were referred to as "targeted areas" and there were about 66 million people living these areas.

In the 2010 Census, several probes were added to the Census questionnaire to help ensure everyone in the households was included on the questionnaire. After listing all the people in the household, respondents were asked a series of questions about the types of people often left off census questionnaires, such as newborns, to make sure someone who should have been on the household roster was not left off accidently.

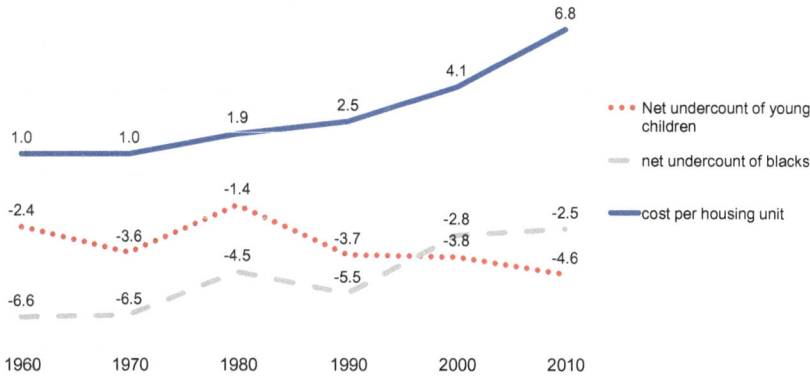

Fig. 14.1 Census cost per household and net undercount rates of young children and blacks: 1960–2010

14.5 Census Costs and Coverage Differentials

The expanded efforts by the Census Bureau to try and eliminate Census undercounts and differential undercounts are reflected in the rising costs of the Census over time. The per household expenditures for the Decennial Census has steadily increased from $16.89 in 1960 to $114.93 in 2010 cost in constant 2009 dollars (National Research Council 2010, p. 32).

The increased cost seems to be related to improvement in the coverage of some groups but not others. Figure 14.1 shows the per household costs along with the net undercount for two hard-to-count groups, namely Blacks and young children. Data for the per household costs is expressed as a ratio of the 1960 value. Data for the undercounts of Blacks and young children were taken from earlier Chapters of this book.

As Census costs rose steeply after 1990, the net undercount of Blacks declined from 5.5% in 1990 to 2.5% in 2010, but the net undercount of young children increased from 3.7% in 1990 to 4.6% in 2010. This suggests that the increased money for outreach has been more focused and/or more effective on some hard-to-count groups than others.

14.6 The Emergence of Philanthropy

One part of the changing landscape regarding Census promotion is increased involvement of philanthropic partners working with the Census Bureau to reach hard-to-count populations. Over the past 20 years, philanthropic foundations and their grantees have become much more involved in the Decennial Censuses. This has evolved into an informal public-private partnership. The foundation/non-profit world mounted efforts in 2000 and 2010 to raise awareness about the importance of the Census and foundations provided funding for many of the "trusted voices" to help increase responsiveness in hard-to-count communities (Michaels 2010; Goldstein 2011; O'Hare 2009). Crews et al. (2011) describe activities of philanthropic foundations in the 2010 Census and set out plans for the 2020 Census. Crews (2011) reports that foundations contributed at least $33 million dollars in Census promotion related to the 2010 Census.

In the 2020 Census, the work of foundations in Census promotion has increased (Daniels 2018; Gupta 2017). In the 2020 Census, more foundations are involved, the work started earlier, and the amount of money supplied by foundations is higher than ten years ago. The work of the philanthropic community has been focused on hard-to-count communities and decreasing the persistent Census undercount differential.

14.7   Summary

Over the past several decades, the Census Bureau has tried many different approaches to reduce or eliminate differential undercounts. Some differential undercounts have decreased in recent decades while others have increased. Some of the efforts have good evaluations to show they improved the count while data for other efforts are less convincing.

References

Barron, W. G. (2001, March 1). *Memo to secretary Donald L. Evans, regarding recommendation on adjustment of census counts*. Washington, DC: U.S. Census Bureau.

Bates, N. (2017). Hard-to-survey populations and the U.S. census: Making use of social marketing campaigns, the Morris Hansen lecture. *Journal of Official Statistics, 33*(4), 873–885.

Choldin, H. M. (1994). *Looking for the last percent: The controversy over census undercounts*. Rutgers University Press.

Crews, K. (2011). *Philanthropic support for the 2010 census outreach: An overview of grants awarded*. Funders Census Initiative. http://www.funderscommittee.org/files/2__Overview_of_Grants_Awarded_by_Kim_Crews-_final.pdf.

Crews, K., Lowenthal, T. A., & O'Hare, W. P. (2011). *Looking to the future: A plan to engage the Philanthropic community in Census 2020 and leverage the benefits of census 2010*. Funders Census Initiative. https://funderscommittee.org/resource/fci-2010-final-report-looking-to-the-future/.

Daniels, A. (2018). Foundations push census turnout in Worrisome Times. *Chronicle of Philanthropy* May 7, 2018.

Darga, K. (1999). *Sampling and the census*. Washington, DC: AEI Press.

Goldstein, W. (2011). *Making Philanthropic dollars count: A history of the funders census initiative during the 2010 Census*. The Hagadorn Foundation. Available online at http://www.funderscommittee.org/files/1__A_History_of_FCI_by_Warren_Goldstein-_final.pdf.

Gupta, V. (2017). Philanthropy and the 2020 census: A once-in-a-decades change to get it right, *National Committee for Responsive Philanthropy, NCRPR'S Quarterly Journal*. Summer 2017.

Hanna, M. G. (2001). Sampling and the politics of representation in US Census 2000. *Environment and Planning, 19*, 515–534.

Hill, R. B. (1975). *Estimating the 1970 census undercount for state and local areas*. Washington, DC: National Urban League.

Media Post Agency Daily. (2017). *Y$R wins a big one: $415 million census 2020 contract*. https://www.mediapost.com/publications/article/283253/yr-wins-a-big-one-415-million-census-2020-contr.html.

Michaels, M. (2010). Grant makers commit millions to help ensure accurate census. *The Chronicle of Philanthropy*, February 21.

National Research Council. (2010) *Envisioning the 2020 census*. Washington, DC: National Academy of Sciences Press.

O'Hare, W. P. (2009). *Testimony before the U.S. House Subcommittee on Information policy, Census and National Archives, Committee on Oversight and Government reform December 2*.

Olson, T. P., Vargas, A., & Williams, J. D. (2014). Mobilizing hard-to-survey population to participate fully in Census and surveys. In R. Tourangeau, B. Edwards, T. P. Johnson, K. M. Wolter, & N. Bates (Eds.), *Hard-to-survey population* (pp. 599–618). Cambridge University Press.

U.S. Census Bureau. (1973). *1970 census of population and housing; evaluation and research program*. The coverage of housing in the 1970 Census. U.S. Census Bureau. https://catalog.hathitrust.org/Record/101681735.

U.S. Census Bureau. (1974). *Effect of special procedures to improve coverage in the 1970 census*. Evaluation and Research Program, PHC (E0-6). U.S. Census Bureau, Washington, DC.

U.S. Census Bureau. (1988). 1980 census of population and housing. Evaluation and research reports. In R. E. Fay. J. S. Passel, J. G. Robinson: with assistance from C. D. Cowan (Eds.), *The coverage of population in the 1980 Census*.

U.S. Census Bureau. (1993). *Programs to Improve Coverage in the 1990 Census,* 1990 CPH-E-3. U.S. Census Bureau, Washington, DC.

U.S. Census Bureau. (2002). A.C.E. Revision II results: Further study of person duplication, DSSD A.C.E. REVISION II MEMORANDUM SERIES #PP-51, Prepared by Thomas Mule, U.S. Census Bureau, Washington, DC.

U.S. Census Bureau. (2003, January 30). *Census 2000 mail response rates: Final report*. Census 2000 Evaluation A.7.a, Herbert F. Stackhouse and Sara Brady, Decennial Statistical Studies Division, U.S. Census Bureau, Washington, DC.

U.S. Census Bureau. (2004a, March). *Coverage improvement in the census 2000 enumeration*. Census 2000 Topic Report No. 10, Census 2000 Testing, Experimentation, and Evaluation Program, TR-10. U.S. Census Bureau, Washington, DC.

U.S. Census Bureau. (2004b, Nov 17). Census 2000: Testing, experimentation, and evaluation program summary results, Florence H. Abramson. U.S. Census Bureau, Washington, DC.

U.S. Census Bureau. (2011). *2010 census: Bilingual questionnaire assessment report*. 2010 Census Planning Memorandum Series, No. 156. U.S. Census Bureau, Washington, DC.

U.S. Census Bureau. (2012a). 2010 census Coverage Measurement Estimation Report: Summary of Estimates of Coverage for Persons in the United States DSSD 2010 Census Coverage Measurement Memorandum Series #2010-G-01. U.S. Census Bureau, Washington, DC.

U.S. Census Bureau. (2012b). *2010 census evaluation of national partnership research report*. 2010 Census Planning Memorandum Series, No. 196. U.S. Census Bureau, Washington, DC.

U.S. Census Bureau. (2012c). *2010 census in schools research final report*. 2010 Census Planning Memorandum Series, No. 179. U.S. Census Bureau, Washington, DC.

U.S. Census Bureau. (2016a). *Developing an Integrated communications strategy*. Select Topics in International Censuses. U.S. Census Bureau, Washington, DC.

U.S. Census Bureau. (2016b). U.S. Census Bureau brings statistics to life for K-12 Classrooms. New Release CB16-146. U.S. Census Bureau, Washington, DC.

U.S. Census Bureau. (2017). *2020 census integrated communication plan*, Version 1.0. U.S. Census Bureau, Washington, DC.

U.S. Census Bureau. (2018). *2020 census complete count committee guide*, D-1280. U.S. Census Bureau, Washington, DC.

U.S. General Accountability Office. (2010a, December). *2010 census: Key efforts to include hard-to-count populations went generally as planned; Improvements could make the efforts more effective for next census*, GAO-11-45. Washington, DC.

U.S. General Accountability Office. (2010b, December 14). *2010 census: Data collection operations were generally completed as planned, but Long-standing challenges suggest need for fundamental reforms*, GAO-11-193. Washington, DC.

U.S. General Accountability Office. (2010c, December 14). *2010 census: Follow-up should reduce coverage errors, but effects on demographic groups need to be determined*, GAO-11-154. Washington, DC.

U.S. General Accountability Office. (2017, October 31). *2020 census; Actions needed to mitigate key risks jeopardizing a cost-effective enumerations*, GAO-18-215t. Washington, DC.

Open Access This chapter is licensed under the terms of the Creative Commons Attribution 4.0 International License (http://creativecommons.org/licenses/by/4.0/), which permits use, sharing, adaptation, distribution and reproduction in any medium or format, as long as you give appropriate credit to the original author(s) and the source, provide a link to the Creative Commons license and indicate if changes were made.

The images or other third party material in this chapter are included in the chapter's Creative Commons license, unless indicated otherwise in a credit line to the material. If material is not included in the chapter's Creative Commons license and your intended use is not permitted by statutory regulation or exceeds the permitted use, you will need to obtain permission directly from the copyright holder.

Chapter 15
Getting Ready for the 2020 Census

Abstract As this book is being written, the 2020 Census is getting closer and the Census Bureau's plans are becoming more concrete. However, there are several factors which raise concerns about whether differential undercounts will be reduced in the 2020 Census. Lack of adequate funding for the Census Bureau and high reliance on the internet are two areas which may lead to higher differential undercounts in 2020.

15.1 Introduction

Up to this point in the book, we have been looking backwards at experiences in previous Censuses. In this Chapter, we will be looking forward to the upcoming 2020 Census. There are several reasons why the 2020 Census will be very different than those that came before it.

Counting every person in the nation—once, just once, and in the right location—is a huge and complex task under any circumstances. The Census Bureau (2017d, p. 6) has documented many of the challenges facing the 2020 Census including:

- Constrained fiscal environment
- Rapidly changing use of technology
- Information explosion
- Distrust in government
- Declining response rates
- Increasingly diverse population
- Informal, complex living arrangements
- A mobile population.

Some of these challenges are on-going, some are new, and some have increased recently. For a number of reasons, the 2020 Census may be the most difficult in our country's history (O'Hare and Lowenthal 2015).

Concerns about the 2020 Census are reflected in the fact that the U.S. General Accountability Office (2017a) put the 2020 Census on its "High-Risk list" in February of 2017. According to U.S. General Accountability Office (2017a, p. 220),

© The Author(s) 2019
W. P. O'Hare, *Differential Undercounts in the U.S. Census*,
SpringerBriefs in Population Studies, https://doi.org/10.1007/978-3-030-10973-8_15

These cost risks, new innovations, and the acquisition and development of IT systems for the 2020 Census, along with other challenges we have identified in recent years, raise serious concerns about the Bureau's ability to conduct a cost-effective enumeration. Based on these concerns, we have concluded that the 2020 Census is a high-risk area and have added it to the High-Risk List in 2017.

In July 2018, the U.S. General Accountability Office (2018) also raised questions about whether the Census Bureau was adequately addressing issues related to hard-to-count populations.

For the 2020 Census, research, planning, and testing are particularly important because the Census Bureau is going to be using several new methods in 2020. According to the 2020 Census operation plan issued in the fall of 2017 (U.S. Census Bureau, 2017d, p. 8) there are four major methodological changes incorporated in the 2020 Census plans:

(1) Reengineering address canvassing—much of the address canvassing will be done in house rather than sending canvassers to walk every street.
(2) Optimizing self-response—respondents will be encouraged to respond through the internet.
(3) Utilizing administrative records and third-party data—administrative records can be used to help determine which housing units are vacant and what people live in housing units that are unresponsive.
(4) Reengineering field operations—enhanced use of technology in communication and administrative tasks can make field operations more efficient.

Together these changes were put in place to save $5 billion compared to what it would cost to repeat the 2010 methodology in 2020. The reduced cost of the 2020 Census was mandated by Congress several years ago. In 2011, Congress told the Census Bureau that the 2020 Census must be conducted for the same cost as the 2010 Census. Language in the 2011 Senate Commerce, Justice, Science Appropriation Bill (U.S. Census Bureau 2012a, p. 2) states, "Therefore, the Committee directs the Bureau to consider budgeting for the 2020 Decennial Census at a level less than the 2010 Census, and to further consider spending less than the 2010 Census, not adjusted for inflation".

In response to this language, the Census Bureau devised a plan for the 2020 Census that would make the cost of the 2020 Census the same as the 2010 Census on a per household basis. Cost containment has always been a concern for Census planning, but more than any recent Census, the 2020 Census operations have been driven by the budget rather than on the methods that would produce the most accurate results. This makes the 2020 Census like no other in recent history.

Despite the effort on the part of the Census Bureau to design a Census that would reduce costs relative to the 2010 Census, congressional funding has still been an issue. For several years prior to Fiscal Year (FY) 2018, the Census Bureau did not receive as much from Congress as it asked for (The Census Project 2018).

In FY2018, the Census budget became a huge issue for Census advocates. The Administration proposed a budget of $1.5 billion for the Census in FY2018 which is essentially the same amount as FY2017. Outside experts said that the Census Bureau needed at least $1.8 billion in FY2018 to prepare for the 2020 Census (Lerner et al 2017, p. 24). In previous decades, year seven to year eight of the Census cycle has seen big budget increases. For example, between 1987 and 1988 it increased by 65%, between 1997 and 1998 it increased by 101%, and between 2007 and 2008 it increased by 61% (Sherman 2017). The increased budget in the few years right before the Census year reflects the need for increased testing and preparation for what has been called the biggest government mobilization outside of war time.

The underfunding of the Census Bureau in FY2018 was a particularly difficult problem. The Bureau had to curtail many tests because of budget shortfalls or budget uncertainty (Vargas 2018). A good example of this is the end-to-end Census tests planned for 2018. For the last several Censuses a thorough, end-to-end test (formerly called dress rehearsal) was vital for finalizing Decennial Census plans. This activity is very important because it is the last chance to test the methods and systems planned for the Census in isolation and collectively. Originally the Census Bureau had planned to do the 2018 end-to-end test in three sites (an area in Washington State, rural West Virginia, and Providence, Rhode Island). Because of budget uncertainty and shortfalls, the end-to-end tests in Washington State and West Virginia were cancelled. In addition, some of the activities originally planned for Providence (such as a media outreach and coverage measurement test) were curtailed. The cuts in 2018 were on top of tests in 2017 on Indian Reservations and in Puerto Rico that had also been cancelled for budgetary reasons.

The lack of rigorous testing that has been the hallmark of past Census planning has put the 2020 Census on a troubling trajectory. The fact that the testing has been reduced at a time when several new methodologies are going to be relied on heavily makes the situation particularly worrisome.

Following an extensive Department of Commerce internal review, in October 2017, Secretary of Commerce Ross (2017) testified in favor of increasing the life cycle budget of the 2020 Census by more than $3 billion and raising the amount the Census Bureau received in FY2018. In the Spring of 2018 (half way through the fiscal year), Congress passed a budget for the Census Bureau with increased funding relative to the original FY2018 budget allocation.

Both the Senate and the House budgeted more for the Census Bureau than the administration asked for in FY2019, but there are still big differences between the two chambers. As this Chapter is being written, the FY2019 budget situation looks more promising than the census budget situation of a year ago, but there is still a lot of uncertainty. It probably goes without saying that trying to plan for something as large and complex as the U.S. Decennial Census when there are shifting, and uncertain funding scenarios makes a difficult task even more challenging.

15.2 Other Issues Hampering 2020 Census Planning

In addition to the budget problems, the Census Bureau and Census advocates have had to deal with several other "distractions" in 2017 and 2018. In June 2017, the Director of the Census Bureau (John Thompson) resigned. The Director of the Census Bureau is a political appointment. At one point, the administration proposed appointing a person to head the Census Bureau who had no experience running a large bureaucracy and no experience with large scale data collection (Mervis 2018a). In addition, his main connection with the Census was through partisan gerrymandering. Advocates pointed out that appointing someone with this kind of political history would hamper the Census Bureau's ability to promote itself as an objective scientific data collection organization and would feed distrust. After advocates questioned the nomination, the Administration withdraw the name from consideration. As this was going on, the Associate Director for the 2020 Census was replaced in the fall of 2017 (Jarmin 2017).

As this Chapter is being finalized in November 2018, the Administration has proposed Dr. Steve Dillingham for the Census Director job. The nominee has held positions directing other federal statistical agencies. We are more than a year and a half into the new Administration and there is no permanent Census Bureau Director in place at a critical point in 2020 Census planning (Mervis 2018b).

In 2018, the Census Bureau received two more surprises from the Administration in terms of the questions to be asked in the 2020 Census. By law, the Census Bureau must submit to Congress the questions it plans to ask in the Census two years before the Census date, thus the questions planned for the 2020 Census were to be submitted by April 1, 2018. On March 26, 2018, Secretary of Commerce Ross (2018) released a memo requiring the Census Bureau to add a question on citizenship to the 2020 Census questionnaire despite the Census Bureau's (Abowd 2018) advice that such a question would diminish the quality of the data and increase the costs of the 2020 Census.

After intensive research on citizenship data, a recent paper issued by the Census Bureau (Brown et al. 2018) concludes, "The evidence in this paper also suggests that adding a citizenship question to the 2020 Census would lead to lower self-response rates in households potentially containing noncitizens, resulting in higher field work costs and a lower-quality population count".

As this book is being written there are six lawsuits challenging the governments' decision to add a question on citizenship status to the 2020 Census questionnaire. Census advocates have mounted a strong campaign to try and get the citizenship question removed. In response to a Federal Register Notice regarding the 2020 Census plans, more than 250,000 individuals and organizations sent in comments opposing the addition of the citizenship question (The Leadership Conference Education Fund 2018).

The concerns about adding a question on citizenship raised by the Census Bureau staff are echoed by six former directors of the Census Bureau (Barabba 2018) who stated, "In summary, we believe that adding a citizenship question to the 2020 census will considerably increase the risks to the 2020 enumeration". The American Statistical Association (2018), the American Sociological Association (2018), the Population Association of America (2018), the American Association of Public Opinion Research (2018), and the Consortium of Social Science Associations (2018) are all on record as opposing the addition of the citizenship question to the 2020 Census. The prestigious National Academy of Sciences, Committee on National Statistics Task Force on the 2020 Census (2018) warns, "According to the Census Bureau's own analysis, addition of the citizenship question could adversely affect the quality and the cost of the 2020 Census".

Aside from the disruptive aspect of adding a question on citizenship, there is another issue about the motive for adding such a question. Ostensibly, the question on citizenship was added to the Decennial Census to provide block level data for the Citizenship Voting Age Population which the Justice Department said it needed to better enforce the Voting Rights Act. It is worth noting that the Voting Rights Act has been in place since 1965, and there has never been block-level Citizen Voting Age Population data available during that period. Data on citizenship is already collected in the ACS and made available at the block group level.

Despite the Commerce Secretary's denial that the addition of a question on citizens was politically motivated, documents that have recently been made available through litigation discovery indicate the addition of the question on citizenship was put on the Census for politically motivated reasons, namely to suppress response to the Census from immigrants (Wines 2018; Bahrampour 2018).

The addition of the citizenship question by itself would probably not be so problematic if it were not for the changing political landscape. The addition of the citizenship question is one step among many that has increased the fear of the federal government in many communities. Over the past year there have been an increasing number of incidents that are likely to discourage the participation in the 2020 Census (Barabba and Flynn 2018). Even before the citizenship question was added, Census Bureau researchers (Meyer and Goerman 2018; U.S. Census Bureau 2017a, b) found respondents less willing to cooperate given the growing climate of fear and mistrust.

For the Census to be successful, respondents must trust the Census Bureau to treat their data confidentially. Trust is likely to be even more important in the 2020 Census. The current political climate and statements from government officials have exacerbated fears in many immigrant communities across the country. Changes in the political climate around immigrants are summarized by the Migration Policy Institute (2018, p. 1) thusly, "the White House has framed immigrants, legal and unauthorized alike, as a threat to American's economic and national security and embraced the idea of making deep cuts to legal immigrations".

The impact goes beyond the undocumented community itself and touches many closely associated groups. According to the National Bureau of Economic Research, (2018, p. 1), "Though not at personal risk of deportation, Hispanic citizens may fear their participation could expose non-citizens in their network to immigration authorities".

Recent decisions by the Trump Administration to expand the "denaturalization" process has expanded the pool of Census respondents who are likely to be leery of giving the federal government information about themselves (Mazzei 2018). Denaturalization is a process where the government can take citizenship away from those who have been naturalized. So, at least a segment of the 21.2 million of foreign-born naturalized citizens (U.S. Census Bureau, American Factfinder, Table S0501) now have reasons to be fearful of giving personal information to the federal government.

A second surprise related to questions in the 2020 Census is linked to how the Census Bureau collects data on race and Hispanic Origin. For many years, the Census Bureau worked on ways to collect better data on race and Hispanic Origin in their data collection activities including the Decennial Census (U.S. Census Bureau 2012b). While the official government categories for race and ethnicity are determined by the U.S. Office of Management and Budget (OMB), the Census Bureau is probably the major user of the categories.

In the government statistical system, race and Hispanic Origin status are two different concepts. Individuals are first asked if they are of Hispanic Origin, then they are asked about their race. One of the biggest problems with the race and Hispanic Origin question used in the 2010 Census is that about half of Hispanics selected the "some other race" racial category because they did not see themselves as belonging to any of the other race categories offered.

In late 2017, the Census Bureau proposed a new set of questions on race and ethnicity to be included in the 2020 Census based on many years of research (U.S. Census Bureau 2017c). However, the U.S. Office of Management and Budget never acted on this recommendation, so as the Census Bureau was getting ready to conduct the end-to-end test Census in April of 2018 they had to fall back on the old race and ethnicity standards in the 2010 Census (U.S. Census Bureau 2018a).

There have been several other distractions with respect to Census planning over the past couple of years. For example, in July 2018, it was announced that the company that had been awarded the contract to print all the material for the 2020 Census (CONVEO) including millions of questionnaires had gone into bankruptcy and the Justice Department had terminated the contract (Wang 2018). It should be mentioned that this printer was selected by the Government Printing Office, not the Census Bureau. It is unclear exactly what impact this will have on the 2020 Census, but it is one more distraction.

These (and other) issues have made it difficult for the Census Bureau and Census advocates to devote full attention to planning the 2020 Census in recent years. The underfunding and distractions related to 2020 Census planning will make it challenging to conduct a high-quality enumeration of everyone in the country.

15.3   The 2020 Census and Differential Undercounts

Given all the methodological changes and new barriers likely to affect the 2020 Census described in this Chapter, one must ask what this all means for undercounts and differential undercounts in the 2020 Census. There are several aspects of the 2020 Census that raise concerns about differential undercounts.

Table 15.1 Mail return rates and omission rates in the 2010 Census

	Mail return rates[a]	Omission rates[b]
Race alone		
White	82.5	4.3
Black	70.0	9.3
American Indian and Alaskan Native	69.8	7.6
Asian	75.4	5.3
Native Hawaiian and Other Pacific Islanders	59.7	7.9
Hispanic	69.7	7.7
Owner-occupied housing units	85.8	3.7
Renter-occupied housing units	66.9	8.5

[a]*Source* Mail Return Rate, U.S. Census Bureau (2012) 2010 Census Mail Response/Return Rates Assessment Report. 2010 Census Planning Memorandum Series, No. 198
[b]*Source* U.S. Census Bureau (2012) 2010 Components of Census Coverage for Race Groups and Hispanic Origin by Age, Sex and Tenure in the United States, DSSD 2010 CENSUS COVERAGE MEASUREMENT MEMORANDUM SERIES #2010-E-51

One of the obvious implications of the new methodologies being planned for the 2020 Census is related to the heavy reliance on the internet. There will be five mailings to each housing unit (i.e. each address in the Master Address File) in the 2020 Census but what is sent to households will vary by location (Fontenot 2018). In the 2020 Census, about 80% of households will receive a letter with a unique identifying number and be asked to complete their Census on the internet. The other 20% (mostly people living in areas without good internet service) will be mailed a paper questionnaire along with information allowing them to respond by internet if they want to. After two reminders, people in the first group who have not completed their Census on the internet, will be sent a paper questionnaire in the 4th mailing.

For all households that do not respond on the internet or by returning a completed paper questionnaire, or by telephone, enumerators will be sent out to complete Non-Response Follow Up (NRFU). Households that require NRFU are less likely to be included in the Census and less likely to provide complete or accurate information (Brown et al. 2018).

Table 15.1 shows mail return rates and omissions rates for eight demographic groups in the 2010 Census. In general, groups that have higher mail return rates have lower omissions rates. For example, the population in owner-occupied housing units have a mail return rates (85.8%) almost twenty percent points higher than the population in renter-occupied housing units (66.9%) and the omissions rate for the population in renter-occupied housing units (8.5%) Is more than twice as high as the population in owner-occupied housing units (3.7%). Similarly, the mail return rate for Non-Hispanic Whites Alone (82.5%) is much higher than those of racial and Hispanic minority groups and the omissions rate of Non-Hispanic Whites alone is much lower than those of minority groups.

Table 15.2 Percent responding to the 2016 ACS by internet by race and Hispanic Origin

	Percent responding by internet
White Alone	42.9
Black alone	23.2
American Indian or Alaska Native Alone	23.5
Asian and Pacific Islander Alone	52.2
Hispanic	25.0

Source Authors Analysis of the Census Bureau's 2016 ACS PUMS file on the IPUMS system at the University of Minnesota

Given the heavy reliance on the internet in the 2020 Census, it is important to note that several recent publications (Pew Research Center 2015; Georgetown Center on Poverty and Inequality 2017; Tomer et al. 2017) show disadvantaged groups in the country typically do not have the same level of access to the internet as more advantaged groups. For example, a report by the U.S. Census Bureau (2018b) shows that 84% of Non-Hispanic Whites households had an internet subscription compared to 73% of Non-Hispanic Black Alone households and 77% of Hispanic households. The same report shows 97% of households with income of $150,000 or more had an internet subscription compared to 59% of households with incomes of $25,000 or less. Given the push to get people to respond by internet, the differential access may exacerbate differential Census participation and differential census undercounts.

Table 15.2 shows the share of major race groups and Hispanics who responded to the 2016 ACS by internet. Note the much lower rates of internet response for Blacks, American Indians, and Hispanics relative to Whites and Asians.

Another concern about the heavy reliance on the internet relates to potential problems in rural areas of the country. O'Hare (2017) shows that the percent of rural residents who do not have internet service at home (21%) is much higher than the percent of urban residents without internet access at home (13%). Even when they have internet available it is often slower in rural areas. A recent study by the Brookings Institution (Tomer et al. 2017, p. 15) shows that a lot of rural areas do not have high-speed internet available. O'Hare (2017) shows many African-Americans in the rural South, Hispanics in the rural Southwest, and Indians and Alaskan Natives living in homeland areas do not have internet access at home and internet availability for poor households in these groups is even lower at around 50%. Impoverished minorities living in rural areas were already a hard-to-count population. In addition, during the Census Bureau's Tribal Consultation with the National Advisory Committee it was reported (Alexander 2017, slide 18) that "Some tribes reported that internet response is currently not a viable option for members..." The Institute for Rural Journalism and Community Issues (2017, p. 1) summarized this issue by stating "the problem with pushing online self-response is that many rural areas lack broadband or any internet service, and those people may be undercounted".

The concern about the lack of good internet access in many parts of rural America increased when the Census Bureau cancelled the end-to-end test in rural West Virginia (the only test in a rural area) scheduled for April 2018. It was cancelled for budgetary reasons. The cancelled end-to-end test planned for rural West Virginia could have revealed any problems with the new data collection plans for the 2020 Census that might be more prevalent in rural areas. Now there will be no end-to-end test in any rural sites which might reveal such problems before the 2020 Census.

When the Census Bureau sent canvassers to the test area in West Virginia in the summer of 2017 to get a complete list of addresses the U.S. General Accountability Office (2017b, p. 9) reported the "Internet connectivity was problematic. There were many deadspots where internet and cellphone service were not available".

Adding an internet response option to the 2020 Census is likely to have different impacts on different groups. A study by the U.S. Census Bureau (2015, Table 2) shows how the addition of an internet option in the American Community Survey in 2013 changed the level of self-response for different groups. Self-response in this context includes either a response by mail or a response using the internet option. Two hard-to-count groups (households with a child under age 5 and Hispanics) experienced a statistically significant increase in self-response rates when the internet option was offered, and three groups (Blacks, respondents over age 65, and respondents without a high school diploma) experienced a statistically significant decrease in self-response rates when the internet option was offered.

To be clear, the Census methods proposed for 2020 call for sending out paper questionnaires in areas where there is little or no high-speed internet available and those who do not self-respond on the internet will also be sent a paper questionnaire. But the paper questionnaire mode of data collection will not be emphasized as much as the internet response will be. Households that do not respond by internet or mail will be visited by a Census enumerator. It is difficult to assess what the emphasis on internet response will mean in terms of differential response and net undercount rates, but there is concern that focusing on internet response may increase the difficulty of getting a complete and accurate Census count for vulnerable groups.

15.4 Use of Administrative Records

Another methodology to be used more extensively in the 2020 Census is the use of administrative or third-party records to supplement the Census. While some administrative records have been used sparingly in past Censuses, the 2020 Census envisions more extensive use compared to the past. Administrative records include information from such sources as income tax filings, social security files, and Medicare files (Rostogi and O'Hara 2012).

Administrative records will be used to help identify vacant housing units and may be used to provide data for individuals when they don't respond or don't provide complete information. At this writing, the exact use of administrative records in the

2020 Census is not clear. The recent addition of the citizenship question has also led to discussion about use of administrative records with respect to that topic (Brown et al. 2018).

Some researchers worry that portions of the population are not adequately reflected in the administrative data. For example, O'Hare (2016) shows that young children are under-represented in administrative records and the use of administrative records in the 2020 Census may exacerbate the high net undercount of this group has experienced in the Census. Other studies show other undercounted groups such as Blacks and Hispanics, are not fully represented in administrative records. Rastogi and O'Hara (2012) found match rates between administrative data and Census records in the 2010 Census higher for White Alone than for Hispanics or Blacks.

Some analysts wonder if administrative records are adequate for covering the hard-to-count population groups like squatters, the homeless, and unauthorized immigrants (Cohn 2016). A statement from a convening of civil rights groups (Urban Institute 2017, p. 11) captures one major concern about the use of administrative records in the Census, "The central civil rights concern about using AdRecs in the Decennial Census is that vulnerable and hard-to-reach subpopulations may be systematically underrepresented by the new procedure".

Given the unevenness in which groups are represented in Administrative Records, depending on how Administrative Records are used, they could increase some of the undercount differentials in the 2020 Census. There is no doubt that using administrative records instead of repeated visits to non-responding households will save money, but it is not clear yet that it will not compromise quality.

15.5   Summary

There are many reasons why the 2020 Census will be different than other recent Censuses. The Census Bureau plans to use several new methods in the 2020 Census largely to meet budget limitations for the 2020 Census placed on the Census by Congress in 2011. The new methodologies that will be used in the 2020 Census have gone through minimal testing, in large part because of funding shortfalls and funding uncertainties at the Census Bureau. In addition to funding problems, many other distractions have made it difficult for the Census Bureau to fully focus on the 2020 Census and these interruptions and disruptions threaten to compromise the quality of the 2020 Census. Some of the new methodologies being introduced in the 2020 Census, designed primarily to save money, could also exacerbate differential undercounts.

References

Abowd, J. M. (2018). *Technical review of the department of justice request to add citizenship status question to the 2020 Census*. Memorandum for Wilbur L. Ross Jr., January 19, 2018, Section B.1.

Alexander, D. (2017). *Briefing on American Indian Alaska Natives, 2020 tribal consultation meetings*. National advisory committee spring meeting. Washington, DC: U.S. Census Bureau, 2017, Slide 18.

American Association of Public Opinion Researchers. (2018). *AAPOR statement regarding 2020 Census*. https://www.aapor.org/Publications-Media/Public-Statements/AAPOR-Statement-Regarding-2020-Census.aspx.

American Sociological Association. (2018). *ASA fights against adding citizenship question to Census*. http://www.asanet.org/census-citizenship-question.

American Statistical Association. (2018). *ASA statement regarding decision to add citizenship question to Decennial Census*. https://www.amstat.org/asa/files/pdfs/POL-CitizenshipQuestion.pdf.

Bahrampour, T. (2018, July 24). Wilbur Ross actively pushed to add citizenship question to 2020 census, documents show. *The Washington Post*.

Barabba, V. (2018). *Letter from Vincent P. Barabba and Six Former directors of the U.S. Census Bureau to Wilbur Ross, Jr., secretary of commerce*. January 26, 2018. Available at https://goo.gl/L3694b.

Barabba, V. & Flynn, K. H. (2018, January 30). Ensure everyone is counted. *U.S. News*.

Brown, J. D., Heggeness, M. L., Dorinski, S. M., Warren, L., & Yi, M. (2018, August 6). *Understanding the quality of alternative citizenship data sources for the 2020 census*.

Cohn, D. (2016, April). *Census Bureau hopes to use data from other government agencies in 2020*. Washington, DC: Pew Research Center.

Consortium of Social Science Associations. (2018, March 27). *COSSA statement on the impact of a citizenship question in the 2020 Decennial Census*.

Fontenot, Jr., A. E. (2018, June 14). *2020 census program update: Presentation to the national advisory committee*. https://www2.Census.gov/cac/nac/meetings/2018-06/fontenot-2020-update.pdf.

Georgetown Center on Poverty and Inequality and Leadership Conference Education Fund. (2017). *Counting everyone in the digital age*. Washington, DC: The Leadership Conference Education Fund.

Jarmin, R. (2017, October 24). *Memo to census staff*.

Lerner, S., Lopez, M., Mayers, T., Lasky, D., Mayerson, D., & Wice, J. (2017). *The count starts now: Taking action to avoid a Census 2020 crisis*. Common Cause New York.

Mazzei, P. (2018, July 23). Congratulations, you are now a U.S. citizen, unless someone decides later you're not. *New York Times*.

Mervis, J. (2018a, May 16). Exclusive: The would-be U.S. census director assails critics of citizenship question. *Science*. https://www.sciencemag.org/news/2018/05/exclusive-would-be-us-Census-director-assails-critics-citizenship-question.

Mervis, J. (2018b, July 20). Census Bureau nominee becomes lightning rod for debate over 2020 Census. *Science*, https://www.sciencemag.org/news/2018/07/census-bureau-nominee-becomes-lightning-rod-debate-over-2020-census.

Meyers, M. & Goerman, P. (2018, May 16–19). Respondents confidentiality concerns in multilingual pretesting studies and possible effects on response rates and data quality for the 2020 Census. In *Paper Delivered at the Annual Conference of the American Association of Public Opinion Research*. Denver CO.

Migration Policy Institute. (2018). *U.S. immigration policy under Trump: Deep changes and lasting impacts*, by S. Pierce, J. Bolter, & A. Selee (Eds.). Washington, DC: MPI.

National Academy of Sciences, Committee on National Statistics Task Force on the 2020 Census. (2018, August 7). *Letter to Jennifer Jessup, department to commerce, committee on national statistics: Task force on the 2020 Census*.

National Bureau of Economic Research. (2018). *Fear and the safety net: Evidence from secure communities*, Marcell Alsan and Crystal Yang, NBER working paper 2473. Cambridge MA: NBER.

O'Hare, W. P. (2016). What do we know about the presence of young children in administrative records. In *Proceedings of the Federal Conference on Statistical Methodology*.

O'Hare, W. P. (2017). *2020 census faces challenges in rural America*. Carsey School of Public Policy, University of New Hampshire, https://carsey.unh.edu/publication/2020-Census.

O'Hare, W. P. & Lowenthal, T. (2015, Fall). The 2020 census: The most difficult in history? *Applied Demography Newsletter*.

Pew Research Center. (2015). *Americans' internet access 2000–2015*. Washington, DC: Andres Perrin Pew Research Center.

Population Association of America. (2018). *PAA statement on citizenship question added to 2020 census*. http://www.populationassociation.org/2018/0/27/paa-statement-on-citizenship-question-for-2020-Census/.

Ross, W. (2017, October 31). *Statement before the Senate Homeland Security and Governmental Affairs Committee*, U.S. Senate.

Ross, W. (2018, May 26). *Reinstatement of a citizenship status question on the 2020 Decennial Census questionnaire*. Memorandum from Secretary of Commerce Wilber Ross to Undersecretary of Commerce Karen Dunn Kelley.

Rostogi, S. & O'Hara, A. (2012). *2010 census match study report*. 2010 Census planning memorandum series, No. 247, November. Washington, DC: U.S. Census Bureau.

Sherman, A. (2017). *Census funding in crisis, center for budget and policy priorities*, Washington, DC. https://www.cbpp.org/blog/census-funding-in-crisis.

The Institute for Rural Journalism and Community Issues. (2017). *The rural blog, rural areas risk being undercounted in the 2020 census*. Downloaded from http://irjci.blogspot.com/2017/10/some-rural-areas-risk-being.html on November 4, 2017.

The Census Project. (2018). *Information from the census project*. Available at https://thecensusproject.org/.

The Leadership Conference Educational Fund. (2018). *Groundswell of opposition tells commerce department to scrap citizenship question on 2020 census*. https://civilrights.org/groundswell-of-opposition-tells-commerce-department-to-scrap-the-citizenship-question-on-2020-census/.

Tomer, A., Kneebone, E., & Shivaram, R. (2017). *Signs of digital distress: Mapping broadband availability and subscription I American neighborhoods*. Washington, DC: The Brookings Institution.

Urban Institute. (2017*). Administrative records in the 2020 U.S. Census: Civil rights consideration and opportunities*. Washington, DC: The Urban Institute.

U.S. Census Bureau. (2012a). *United States Census 2020: FY2012 appropriation—120 day report to congress*. U.S. Census Bureau: Washington, DC.

U.S. Census Bureau. (2012b). *2010 Census Race and Hispanic origin alternative questionnaire experiment*, 2010 Census planning memorandum series no. 211, August.

U.S. Census Bureau. (2015). *An examination of self-response for hard-to-interview groups when offered an internet reporting option for the american community survey*. 2015 American Community Survey Research and Evaluation Report Memorandum Series ACS15-RER-10. Washington, DC: U.S. Census Bureau.

U.S. Census Bureau. (2017a). *Respondent confidentiality concerns*. Memorandum for Associate Directorate for Research and Methodology (ADRM) from Center for Survey Measurement (SCM), September 20, 2017.

U.S. Census Bureau. (2017b). *Respondent confidentiality concerns and possible effects on response rates and data quality for the 2020 census*. Presentation by Mikelyn Meyers at the Census Bureau National Advisory Committee Meeting, November 2, 2017.

U.S. Census Bureau. (2017c, February). *2015 national content test race and ethnicity analysis report, a new design for the 21st century, version 1.0*. Washington, DC: U.S. Census Bureau.

U.S. Census Bureau. (2017d). *2020 census operational plan: A new design for the 21st century, version 3.0*. Washington, DC: U.S. Census Bureau.

U.S. Census Bureau. (2018a). Memo from Albert E. Fontenot Jr., to the record; using two separate questions for race and ethnicity in 2018 end-to-end census test and 2020 census. Dated January 26, 2020 Census Program Memorandum Series 2018.02. Washington, DC: U.S. Census Bureau.

U.S. Census Bureau. (2018b). *Computer and internet use in the United States: 2016. American community survey reports, by Camille Ryan, ACS-39*, Issued August.

U.S. General Accountability Office. (2017a, February 15). *High risk series: Progress on many high-risk areas, while substantial efforts needed on others, GAO-17-317*.

U.S. General Accountability Office. (2017b, October 31). *2020 Census, actions needed to mitigate key risks jeopardizing a cost-effective enumeration, GAO-18-215T*. Testimony by Gene L. Dodaro, Before the Committee on Homeland Security and Government Affairs, U.S. Senate.

U.S. General Accountability Office. (2018, July 2018). *2020 Census, actions needed to address challenges to enumerating hard-to-count groups, GAO-18-599*.

Vargas, A. (2018). *2020 census program update: Discussant presentation*, U.S. Census Bureau National Advisory Committee meeting, June 14. Washington, DC: U.S. Census Bureau.

Wang, H. L. (2018). *Officials botched 2020 Census printing contract, report finds*. July 31 National Public Radio.

Wines, M. (2018, July 10). Why was a citizenship question put on the Census? "Bad faith", A judge suggests". *New York Times*.

Open Access This chapter is licensed under the terms of the Creative Commons Attribution 4.0 International License (http://creativecommons.org/licenses/by/4.0/), which permits use, sharing, adaptation, distribution and reproduction in any medium or format, as long as you give appropriate credit to the original author(s) and the source, provide a link to the Creative Commons license and indicate if changes were made.

The images or other third party material in this chapter are included in the chapter's Creative Commons license, unless indicated otherwise in a credit line to the material. If material is not included in the chapter's Creative Commons license and your intended use is not permitted by statutory regulation or exceeds the permitted use, you will need to obtain permission directly from the copyright holder.

Chapter 16
Summary

Abstract Differential undercounts in the US Census are a substantial and on-going problem. Some of the largest undercount differentials are outlined in this Chapter and the relationship between net undercounts and omissions is reiterated. The importance of the Census is highlighted and ways in which readers can get involved in promoting a better 2020 Census are described here.

16.1 Introduction

The big picture regarding the accuracy of the US Census since 1950 is a good news/bad news story. Over the past 60 years, the overall accuracy of the US Decennial Census has steadily improved, but some groups have persistently experienced higher net undercounts than other groups in the Census and some new differential undercounts (for example, young children compared to adults) have emerged over this period.

There is also a good news/bad news story associated with the 2010 Census coverage as well. The good news is, the coverage error for the total population was very small by international and historic standards, but the bad news is there are still large coverage differences among groups. Data presented in this book underscore the extent to which some groups have higher net undercount and omissions rates than others in the Census.

Who is missing? Based on the groups for which the Census Bureau provides net undercount and omissions data, there are a handful of groups that are at exceptionally high risk of being missed in the Census. To put the data below in context, the Non-Hispanic White Alone population had a net overcount of 0.8% and an omissions rate of 3.8% in the 2010 Census The groups listed below are those with the highest net undercount and omissions rates based on Census Bureau data:

- Young children (age 0–4) had a higher net undercount and higher omissions rate than any other age group in the 2010 Census. In the 2010 Census, the net undercount rate for young children was 4.6% and the omissions rate were 10.3%. The net undercount rate for young children has increased rapidly since the 1980 Census.

© The Author(s) 2019
W. P. O'Hare, *Differential Undercounts in the U.S. Census*,
SpringerBriefs in Population Studies, https://doi.org/10.1007/978-3-030-10973-8_16

- The Black population had a higher net undercount than any other race/Hispanic group in the 2010 Census. The net undercount rate for the Black population was 2.1% (2.5% based on the Demographic Analysis method) and the omissions rate was 9.3%. Black males age 20–60 have exceptionally high net undercount and omissions rates. The net undercount rate for Black males age 30–49 was 10% and the omissions rate for this group was 16.7%. Historically, the Black population experienced high net undercount rates in the Census. While the net undercount of Blacks has decreased over time, it is still relatively high and the differential undercount between Blacks and Nonblacks has improved little since 1940.
- Hispanics had a net undercount rate of 1.5% in the 2010 Census and an omissions rate of 7.7%. Hispanic males age 20–50 had very high net undercount and omissions rates. The net undercount rate for Hispanic males age 30–49 was 5.1% and there was an omissions rate of 10.9% in the 2010 Census.
- In the 2010 Census the net undercount rate for American Indians living on reservations was very high at 4.9%.

One of the themes that flows through the various groups that are most likely to be missed in the Census is social and economic marginalization. Higher income groups like Non-Hispanic Whites are counted accurately or have over counts, while less affluent groups like Blacks, Hispanics, and American Indians on reservations have significant undercounts. In addition, young children, who have a high net undercount, are politically powerless. These groups are often under-represented in terms of political power and/or civic participation as well. In some ways, the Census undercounts of these groups is just one more way they are sidelined and under-represented in society. Given the connection between Census counts and federal funding, the groups most in need of federal assistance are the groups least likely to get their fair share because they have the highest net undercounts and omissions.

16.2 Net Undercounts and Omissions

Another point that was made repeatedly in this book is the fact that net undercounts and omissions are not the same thing. In some cases, a low net undercount rate for a group might lead people to assume no one was missed in the group but that is often not the case. For example, the net undercount for Asians was zero in the 2020 Census, but Asians had a 5.3% omissions rate which is somewhat higher than the rate for Non-Hispanic White Alone (3.8%).

There was a very small net undercount for American Indians and Alaskan Natives Alone or in Combination in 2010 but the omissions rate for American Indians and Alaskan Natives was 7.6% which is double the rate for Non-Hispanic White Alone (3.8%). The count of Native Hawaiian and Pacific Islanders in the U.S. Census is relatively accurate but the omissions rate for Native Hawaiian and Pacific Islanders (7.9%) is about double the rate for Non-Hispanic White Alone (3.8%).

Table 16.1 Undercount rates for individual factors and a combination of factors

	Net undercount
Blacks	−2.5
Males	−0.8
Age 30–49	−0.8
Renters	−1.1
Black male renters age 30–49	−12.2

Source Taken from various chapters in this book

16.3 Cumulative Impact

One point woven into the analysis in previous Chapters is how the accumulation of risk factors has a multiplicative impact on net undercount rates. Table 16.1 shows net undercount rates for four individual factors (age, race, sex, and tenure) and then shows the net undercount rate when all of these factors are combined. None of the net undercount rates for individual factors is more than 2.5%, but collectively the net undercount rate for someone with all four factors is 12.2%. If we had more data, I strongly believe the addition of other factors like poverty, language ability, and household structure would further drive up the net undercount rate for people with multiple risk factors.

16.4 The 2020 Census

It is worth repeating that Census data are the backbone of our democratic system of government. The significance of the Census envisioned by the founding fathers is reflected in the fact that Decennial Census is mentioned in the sixth sentence of the Constitution and is the first responsibility given the new federal government.

Chapter 2 provides detailed information about the many ways the Decennial Census data are used. Census-related figures are used to distribute more than $850 billion in federal funding each year to states and localities. Data from the 2020 Census will guide the distribution of roughly $25 trillion in federal assistance to states and localities between 2021 and 2030. Countless decisions in the public and private sectors are based on Census data. Moreover, the impact of flaws in the census counts often last a decade because population estimates, projections, and survey weights, are derived from Census counts

Chapter 15 in this book outlines some of the difficulties that have hindered 2020 Census planning. But much of what will determine the accuracy of the 2020 Census will depend on what happens after this book is published.

The accuracy of the 2020 Census will depend on the balance between two broad set of forces. Many factors like lower response rates to surveys, growing distrust of the federal government, dependence on new methodologies, and underfunding of the 2020 Census would lead one to expect the accuracy of the 2020 Census will be worse than in 2010 and differential undercounts will increase. But the mobilization of groups outside the Census Bureau is more advanced than ever before. Growing numbers of organizations are recognizing the importance of the Census for their work and for the well-being of their community.

I believe the results of the 2020 Census will largely be determined by the balance between these two sets of positive and negative forces. To the extent that advocates, grass roots leaders, elected officials, and others mobilize their communities to be counted, the results of the 2020 Census can be better than 2010 with respect to the net undercounts and omissions in hard-to-count populations and communities. But convincing people to respond to the Census given the current political climate will not be easy. To the extent that the Administration and Congress soften the negative rhetoric and policies aimed at immigrant and minority communities it will be easier to convince people to respond to the Census.

16.5 What Can You Do?

This book has largely been a scholarly research effort. But I want to end with a few notes on how readers can easily get involved in making the 2020 Census a success.

Individuals or organizations can get involved as a Census Bureau Partner by signing up at www.census.gov/partners. Partners will get updated information from the Census Bureau and will receive advice about census-promotion activities. Through this mechanism one can get involved in national, state, and local activities to promote the 2020 Census. One example of the help provided by the Census Bureau, is the community outreach toolkit that is available from the Census Bureau at https://www.census.gov/partners/toolkit.pdf.

One can also join (or start) a state or local Census Complete Count Committee. Complete Count Committees are made up of volunteers and are established by tribal, state, and local governments and community leaders or organizations to increase awareness and motivate residents to respond to the 2020 Census. To learn more about Complete Count Committees or find one near you, contact the Complete Count Committee Program at the Census Bureau at https://www.census.gov/programs-surveys/decennial-census/2020-census/complete_count.html.

One of the best ways to stay up-to-date on Census-related developments in Washington, is to sign up for regular alerts at foundation-funded Census Project. One can also visit the Census Project website to find information on past census-related events https://thecensusproject.org/.

Another site with a lot of good information on hard-to-count populations is The Leadership Conference Education Fund. Their website is https://civilrights.org/census/. Material on the Leadership Conference Education Fund website includes topical briefs and data on a lot of hard-to-count populations.

For those interested in making sure all children are counted accurately in the 2020 Census visit the CountAllkids Complete Count Committee website (www.countallkids.org) or the KIDS COUNT website at the Annie, E. Casey Foundation (www.kidscount.org).

Open Access This chapter is licensed under the terms of the Creative Commons Attribution 4.0 International License (http://creativecommons.org/licenses/by/4.0/), which permits use, sharing, adaptation, distribution and reproduction in any medium or format, as long as you give appropriate credit to the original author(s) and the source, provide a link to the Creative Commons license and indicate if changes were made.

The images or other third party material in this chapter are included in the chapter's Creative Commons license, unless indicated otherwise in a credit line to the material. If material is not included in the chapter's Creative Commons license and your intended use is not permitted by statutory regulation or exceeds the permitted use, you will need to obtain permission directly from the copyright holder.